CHINA'S ENERGY FUTURE

Significant Issues Series
Timely books presenting current CSIS research and analysis of interest to the academic, business, government, and policy communities.
Managing Editor: Roberta Howard Fauriol

The Center for Strategic and International Studies (CSIS) is a nonprofit, bipartisan public policy organization established in 1962 to provide strategic insights and practical policy solutions to decisionmakers concerned with global security. Over the years, it has grown to be one of the largest organizations of its kind, with a staff of some 200 employees, including more than 120 analysts working to address the changing dynamics of international security across the globe.

CSIS is organized around three broad program areas, which together enable it to offer truly integrated insights and solutions to the challenges of global security. First, CSIS addresses the new drivers of global security, with programs on the international financial and economic system, foreign assistance, energy security, technology, biotechnology, demographic change, the HIV/AIDS pandemic, and governance. Second, CSIS also possesses one of America's most comprehensive programs on U.S. and international security, proposing reforms to U.S. defense organization, policy, force structure, and its industrial and technology base and offering solutions to the challenges of proliferation, transnational terrorism, homeland security, and post-conflict reconstruction. Third, CSIS is the only institution of its kind with resident experts on all the world's major populated geographic regions.

CSIS was founded four decades ago by David M. Abshire and Admiral Arleigh Burke. Former U.S. senator Sam Nunn became chairman of the CSIS Board of Trustees in 1999, and since April 2000, John J. Hamre has led CSIS as president and chief executive officer.

Headquartered in downtown Washington, D.C., CSIS is a private, tax-exempt, 501(c) 3 institution.

The CSIS Press
Center for Strategic and International Studies
1800 K Street, N.W., Washington, D.C. 20006
Tel: (202) 887-0200 Fax: (202) 775-3199
E-mail : books@csis.org Web: www.csis.org

CHINA'S ENERGY FUTURE

The Middle Kingdom Seeks Its Place in the Sun

ROBERT E. EBEL

FOREWORD BY JAMES SCHLESINGER

THE CSIS PRESS

Center for Strategic
and International Studies
Washington, D.C.

Significant Issues Series, Volume 27, Number 6

Printed on recycled paper in the United States of America
Cover design by Robert L. Wiser, Silver Spring, Md.
Cover photograph: Farmer watching Three Gorges Dam construction
© Karen Su/CORBIS

09 08 07 06 05   5 4 3 2 1

ISSN 0736-7136
ISBN 0-89206-473-0

Library of Congress Cataloging-in-publication Data
Ebel, Robert E.
China's energy future : the Middle Kingdom seeks its place in the sun / Robert E. Ebel.
p. cm. — (Significant issues series ; v. 27. no. 6)
Includes bibliographical references and index.
ISBN 0-89206-473-0
1. Energy policy—China. 2. Petroleum industry and trade—China.
I. Title. II. Series.
HD9502.C6E24 2005
333.790951—dc22   2005026861

CONTENTS

LIST OF TABLES

FOREWORD

James Schlesinger

Oil is in the headlines. China is in the headlines. The combination of the two has recently become economically and politically explosive. To say, as Robert Ebel disarmingly puts it in his subtitle, that the Middle Kingdom is seeking its place in the sun is something of an understatement. China has already achieved its place in the *energy* sun. It has now become the world's second largest consumer of oil, following the United States. Disappointed in its ability to discover and develop its own domestic reserves, it must now import almost half of its oil supply from abroad—with all that that implies for China's international trade, investment, and security of supply. Both consumption and imports seem likely to grow, as the Chinese economy continues to expand rapidly.

In addition, China is the world's largest producer and consumer of coal—burning some 2 billion tons a year. In the next 20 years, it is projected to burn roughly an additional 1 billion tons. This, of course, has implications for the release of greenhouse gases, about which much concern has been expressed worldwide. Moreover, most of China's coal is burned without pollution-controlled equipment, which has had a major impact not only on China's air quality, but also on the wind-blown transport of pollutants elsewhere, including to the United States. China is also planning a dramatic increase in its use of nuclear power—in part to limit the air-quality consequences of its exploding demand for electricity.

James Schlesinger is the former U.S. secretary of energy.

It is not hard to explain the surging interest in energy matters, which has now generated more public concern (and printers' ink) than it has in several decades. Since the collapse of oil prices in the mid-1980s, public and political interest in oil has receded. But now there is a seriously strained production capacity, and OPEC itself is striving to hold down or push down crude oil prices (or to cash in on the present high price).

For the recent surge in oil prices, China has been held out as a major culprit, given its rapid growth in oil consumption and its rapidly rising dependency on the international oil market. (Elsewhere in the world, the United States is held out as a principal culprit.) Given China's recent immersion in the world oil market, not to mention its vastly expanded dependence on international trade generally, it may be hard to remember that just a generation ago, under Mao Tse-Tung, the watchword in China was self-sufficiency. Mao advocated a stripped autarky, no place more importantly than in energy self-sufficiency.

In 1978, even before the normalization of U.S.-Chinese relations, as U.S. energy secretary I visited Daqing, then very much the dominant center of China's oil industry. (Among other things, I remember how bitterly cold it was.) It was noteworthy, despite the relative backwardness of its extraction technology, that China to that point had been satisfied to rely on its own technology and not to seek technical expertise from abroad. All that has now changed. Daqing itself is in substantial decline. Moreover, since 1978 China has sought the introduction of foreign technology and of international oil companies. And now, given its growing oil dependency, it seeks increasingly to find—and to control—sources of supply abroad.

We should all be grateful to Robert Ebel—and to the Center for Strategic and International Studies—for providing this brief but comprehensive overview of China's energy scene. Robert Ebel is one of those thorough and painstaking students of the international energy scene in general, and of oil in particular. It is widely recognized that China's energy statistics are less than perfect—and sometimes shrouded in mystery. To separate out the chaff and effectively reconcile some of the data has been a real challenge. Ebel has successfully met that challenge. He has provided not only a useful primer on energy development in China (and beyond), but he has also established a credible data baseline for future analyses of the ever-changing Chinese energy scene.

CHAPTER ONE

INTRODUCTION

"If a person wishes to have luck, longevity, health and peace, he or she must live in a world of light."[1]

So began a new Standard Oil advertisement in China from the early twentieth century—not to sell kerosene, one of the major petroleum products the company had to offer, but rather to create a market for this kerosene through the introduction of kerosene lamps. The lamp being advertised was small, made of tin, with a chimney and a wick. But the lamps were not sold; they were given away to the Chinese people. Sales of kerosene to fuel these lamps were imminently successful.

This marketing approach is still applied today, and quite successfully. Now it also applies to cell phones, literally given away, but with users charged a monthly fee. At last count there were 354 million cell phones in China, allowing that country to bypass the time-consuming and expensive construction of landlines and bringing the users headlong into the twenty-first century. Many energy companies seek to do the same: capitalize on the enormous market potential they see in China and create a market where none existed earlier.

China is second only to the United States in consumption of oil and electric power, and its coal industry leads the world in terms of annual output. Expansion of the country's nuclear power industry, which calls for the construction of at least 28 new reactors by the year 2020, has attracted the attention of vendors worldwide. Natural gas requirements, which are surging, will be covered in large part by imports delivered by pipeline and in the form of liquefied natural gas (LNG).

China's burgeoning oil demand and the necessity that this growth can be satisfied only through imports have made China a defining force in the world oil market. A number of benchmarks track China's emergence into the world oil market.

- By the mid-1980s China had become the sixth largest crude oil producer in the world.
- Crude oil exports had hit a peak of 712,000 barrels per day (b/d) in 1985.
- Failures to develop new producing capacity, plus unanticipated demand growth, in the early 1990s transformed China from a "giver" of oil to the world oil market to a "taker."
- In 2004 China exported 110,000 b/d of crude oil and 230,000 b/d of petroleum products, against imports of 2.45 million b/d of crude and 760,000 b/d of products.
- China's oil consumption averaged just 2.23 million b/d in 1990, but more than doubled by 2000, for an increment of roughly 2.6 million b/d. It is quite probable that in 2005 consumption will approach, if not exceed, 6.8 million b/d.

The growth in the gross domestic product (GDP) of the country and the newly acquired capability of the public to enjoy a much higher lifestyle, including automobile ownership, have pushed oil demand upward. Because domestic oil supply is relatively static, reliance on foreign oil is rising. This increasing reliance, in the judgment of most oil market watchers, has been a key factor behind the world oil price increases during 2004 and into 2005.

China's demand is only partly to blame for high oil prices. To begin, the growth in China's oil demand during 2004, as a share of world oil demand growth, had not been much higher than the share held during the preceding decade. Oil-exporting countries gradually had worked off their spare producing capacity during the past decade or so, to the point where the ability to respond quickly to higher world oil demand has essentially disappeared. The 2004 surprise of the measure of China's need for foreign oil might have been covered just by bringing that spare capacity into play, presuming of course the willingness to do so.

Absent corresponding spare producing capacity, pressure on available supplies increased and prices rose accordingly.

HOW CHINA IS RESPONDING

Oil import diversification, acquisition of equity oil abroad, plus expanded development of hydroelectric and nuclear power to reduce dependence on coal, together with an emphasis on energy conservation, define China's energy future. These measures are designed to come together to reduce future oil import dependency to considerably less than the 40 percent marking 2004.

China has sought opportunities worldwide to develop equity oil—that is, oil actually owned by China. Yet, emphasis on equity oil is misguided and incorrect. Equity oil will not protect China's national interests in times of market supply disruption or against high prices. Moreover, China's willingness to overpay for access, as it has, is disruptive in itself. The effort to date has not been successful. Volumes of equity oil secured to date are but a fraction of annual oil imports.

Most of the oil that China imports transits through the Strait of Malacca. The vulnerability of the strait to accidents or increasingly to piracy has moved China to take military-oriented steps to protect passage through the strait. Denial of passage, deriving perhaps out of a conflict with Taiwan, drives military planning.

This military planning has resulted in a broad Chinese military buildup, a buildup reportedly much higher than officially admitted. Indeed, China's military budget ranks highest in Asia and third in the world.[2]

OTHER FORMS OF PRIMARY ENERGY DO MATTER

Coal is still dominant for electric power generation in China, with supplies in 2005 to reach 2 billion tons at least. The coal sector continues to be marked by the deaths of thousands of miners every year, as safety regulations are routinely ignored. Electricity generation is the leading coal consumer, but power shortages have prevailed in recent years, largely because generation cannot keep pace with growth in demand.

The future of nuclear power in China appears secure. If schedules for reactor construction are met, nuclear power capacity in 2020 will be more than five times 2004 capacity. Even so, nuclear power capacity that year would provide just 4 percent of total national generating capacity.

Imports of natural gas by pipeline and in the form of LNG will largely determine those volumes available for domestic use, given the limitations of the country's natural gas resource base. Chinese officials have placed natural gas demand as falling within a range of 70 billion to 112 billion cubic meters (bcm) by 2010, more than doubling to 150 to 250 bcm by the year 2020.

Finally, China calls for consumption of renewables to account for 10 percent of total national energy consumption by 2020, which is more than most forecasts predict. China looks to the West for technology transfers to allow that goal to be met.

IS CHINA'S GROWTH SUSTAINABLE?

China understandably finds it difficult to measure correctly economic activity in a country of 1.3 billion inhabitants. How then to judge the future—a future that to a considerable degree will likely shape, and be shaped by, access to secure sources of oil, natural gas, and other forms of energy?

Rising Chinese oil demand will be accompanied by the need for more and more foreign oil to fill the gap, as growth in domestic crude oil production is questionable. These factors, coupled with minimal spare producing capacity worldwide, will continue to influence prices during the next several years, barring any unforeseen developments in demand or supply.

Yet, continued economic expansion and concurrent growth in oil imports is not necessarily a given. Is past growth sustainable? In addition to current and future oil and natural gas imports, China today has a voracious appetite for steel, copper, and cement, among other commodities.

In sum, China imports so that it can export. Economic expansion in China essentially reflects, and will continue to reflect, its ability to export goods to world markets and the ongoing world demand for those goods. Today it is textiles, tomorrow it will be automobiles, as China puts its natural advantage—low-cost labor and an artificially undervalued local currency—to work. Will world markets always be accepting of Chinese exports? Clearly, there are and will continue to be objections to how China uses these advantages. Any slowdown in Chinese exports, for whatever the reason, likely would be reflected in reduced commodity imports. Countries that have thrived on selling to China would not be insulated from that slowdown. This prospect adds uncertainty to the world oil market and in turn supports continued price volatility.

Data transparency is a growing issue for those trying to understand China's rising energy demand. It is also a concern for companies looking to invest as well as for the Chinese government. Despite the difficulty of tracking the energy-demand habits of 1.3 billion consumers, the government needs to step up to meet international reporting standards for demand, production, imports, and exports.

FUTURE GROWTH IS NOT A GIVEN

China currently confronts a number of problems that could affect future growth trends:

- Growing dependence on imported oil and, soon after, on imported natural gas,
- Mass migration from rural to urban areas,
- Increasing water shortage,
- Doubts about China's ability to feed itself, and
- A population aging so rapidly that China could grow old before it grows rich.

How China responds to these issues may determine the type and degree of growth that it can sustain. In addition, China is struggling to find the right combination of market incentives, conservation, pollution abatement efforts, and energy supply diversification, including diversification among foreign suppliers of oil and gas, that could lead to a reduction in oil import reliance, guaranteeing adequate energy supplies while protecting the national security of the country.

In other words, China is acting no different than other oil-importing countries concerned about the linkage between national security and energy security.

Notes

[1] Exxon, *The Lamp,* Fall 1999, http://www.exxonmobil.com.

[2] Victor Mallet, "Rumsfeld Questions China on Missiles," *Financial Times,* June 5, 2005.

CHAPTER TWO

BACKGROUND

HISTORICAL REVIEW

China's production of crude oil was insignificant before the Communist takeover of 1949. The new government inherited three fields that had an aggregate output of less than 2,000 b/d. None of these fields had the potential to become large producers.

In the 1950s, as today, coal had provided the larger part of the country's energy needs. The limited requirements for liquid fuels and lubricants were supplied mostly from the Soviet Union. Imports rose to a peak of about 60,000 b/d in 1960. But with the opening of the Daqing oilfield that year, imports declined rapidly. By 1963, Daqing was providing 88,600 b/d, which allowed Peking to claim self-sufficiency in oil.

Opening of the Shengli oilfield in 1962 and the Takang field in 1967 followed, and the three fields together accounted for 80 percent of China's crude oil production in 1975. Development of these fields allowed oil production to rise rapidly during the 10-year period 1965–1975 (see table 2.1), to almost 1.5 million b/d and further to 1.67 million b/d in 1976.[1]

World oil had given little attention to China's oil sector until the 1973 Arab oil embargo, which had coincided with China's first commercial sale of crude oil. Always searching for a substitute for Persian Gulf oil, uninformed observers thought that China could become a major player in the market, and one prediction even presented China as a future Saudi Arabia.

The Central Intelligence Agency (CIA), recognizing the scarcity and unreliability of data, was more realistic about the promise of

Table 2.1
Production of Crude Oil in China, Selected Years, 1950–1975

Year	Thousand Barrels/Day
1950	4
1955	19
1960	102
1965	219
1970	564
1975	1,485

Source: Central Intelligence Agency, "China Oil Production Prospects," ER 77-10030U, June 1977, figure 7, p. 9.

Table 2.2
China's Exports of Crude Oil and Petroleum Products, 1980–1985

Year	Thousand Barrels/Day
1980	350
1981	367
1982	409
1983	406
1984	553
1985	712

Source: Fereidun Fesharaki, "China's Downstream Petroleum Industry: Recent Developments and Future Potential," paper presented at the China Conference, CSIS, Georgetown University, Washington, D.C., April 3, 1986, table 3, p. 15.

Chinese oil. The CIA at that time (1977) thought that China would be producing 2.4 million to 2.8 million b/d by 1980, of which only 200,000 to 600,000 b/d would be available for export. Moreover, within a decade or so, continuously expanding domestic demand would absorb total capacity unless deposits in the West or offshore were proved and exploited much more rapidly than expected.

By 1993, reasonably within the bounds of this forecast, China shifted from its position as a net exporter to net importer, for two reasons—rapid growth in domestic demand and stagnation in oil production.

A subsequent report, released in August 1979, observed that China was unlikely to become a major supplier of crude oil to the world in the next decade; although it produced 2 million b/d in 1978, some 90 percent went to cover domestic needs.[2] Moreover, oil production growth was slowing, reflecting political disorders and technical difficulties encountered at larger fields. Because of these constraints, it was believed that exports might level off at about 300,000 b/d in 1982 or so.

How did the forecasts of China as an oil exporter hold up over time? Fereidun Fesharaki, in a paper presented in April 1986, noted that except in the years of declining oil production (1980–1982), crude oil output had consistently exceeded domestic refining capacity, reflecting a rather consistent policy of promoting petroleum exports as a foreign exchange earner. From 1982 on, an average 80 percent of incremental production had been exported.[3] China had become the largest oil exporter in Asia, having surpassed Indonesia in 1983 (see table 2.2).

Crude oil made up more than 80 percent of China's oil exports. Japan was the largest buyer of Chinese crude oil, followed by Singapore and Brazil. Very small volumes found their way to the United States. China, like any oil-exporting country, sought to diversify its export markets and offered considerable implicit discounts to U.S. buyers.

China continued to follow a policy of constraining domestic consumption as a way of maximizing exports and foreign exchange earnings.[4] In the mid-1980s many analysts thought that China would be able to meet its 1990 oil production goal of 3 million b/d, but that the year 2000 goal of 4 million b/d would be more difficult to reach, largely because of declining production onshore. Even so, Fesharaki had projected that for the remainder of the 1980s exports would not surpass 800,000 b/d before beginning to decline in the 1990s and that exports might fall to around 500,000 b/d by 1995.[5]

In the mid-1980s China was the sixth largest producer of crude oil in the world and held the seventh spot in terms of refining capacity. But rapidly expanding domestic requirements, coupled with difficulties in finding and developing large new oil fields, made it quite clear, even then, that China's future as an oil exporter was limited. Yet, most observers failed to identify just how quickly those volumes surplus to domestic requirements would disappear, with that failure reflecting demand estimates that were too low. Forecasts of crude oil production also had been conservative, but with good reason, and were generally accurate in content.

In 1990 oil product consumption in China averaged 2.2 million b/d, compared with 1.6 million b/d in 1980, 10 years earlier, underscoring a steady but not necessarily spectacular growth. In contrast, just five years later, in 1995, consumption had risen to almost 3.4 million b/d and by 2000 was averaging 4.8 million b/d while showing no signs of slackening.

The domestic oil-producing industry simply was not able to keep pace with the rapid growth in demand for oil. Failures to develop major new sources of domestic supply, plus stagnation and decline at producing fields, precluded any significant increases in crude oil production (table 2.3).

In the very early years of the 1990s, supply had continued to exceed domestic needs. For example, in 1990 China's exports—480,000 b/d of crude oil and 108,000 b/d of petroleum products—far exceeded its imports of just 58,000 b/d of crude oil and 62,000 b/d of petroleum products.[6] Increasing demand (see table 2.4) and comparatively stagnant domestic supplies led to an unavoidable consequence: China became a net oil importer of petroleum products in 1992 and of crude oil in 1993, thus shifting away from its role as a "giver" to the world oil market and becoming a "taker," inheriting all the political and financial implications accompanying this shift.

By 1997, imports of crude oil had risen to 706,000 b/d and that of petroleum products to 464,000 b/d. Surprisingly, this growing reliance on imported oil did not translate into the disappearance of oil exports, as might have been expected. Although total oil exports did decline, this decline was limited. Indeed, in 1997 China still held onto its position as a regionally important supplier, marketing abroad 397,000 b/d of crude oil and 111,000 b/d of petroleum products, with product sales actually exceeding those registered in 1990.

At the same time, China was expanding its exploration and development activities outside the country, in the Middle East, Central Asia, and Latin America, in search of equity oil. China early on had recognized that its reliance on foreign oil was not only going to continue but would be expanding in the coming years. As with all oil-importing countries, security of supply through diversity of supply was, and remains today, the driving philosophy behind this search.

In 1998, the government reorganized the petroleum industry, merging most state-owned assets into three main companies—China

Table 2.3
Production of Crude Oil in China, Selected Years, 1990–2005

Year	Million Barrels/Day
1990	2.77
1995	2.99
2000	3.25
2001	3.30
2002	3.39
2003	3.41[a]
2004	3.49[b]
2005	3.52[c]

Source: Unless otherwise noted, from the U.S. Department of Energy, "An Energy Overview of the People's Republic of China," http://www.fossil.energy.gov/international/EastAsia_and_Oceania/chinover.html, 2005.

[a] Feng Xue, "Offsetting Demand on World Supplies," *Petroleum Review,* October 2004, pp. 26–27.

[b] "China Processes Record Crude," *Oil and Gas News* 23, no.3 (January 17–23, 2005), http://www.oilandgasnewsonline.com. Crude production in 2004 was reported to be 2.9 percent above the 2003 level. See also "China's Oil Giant Thrives on Record Crude Prices," *Business Report 2005,* January 14, 2005, http://www.busrep.co.za.

[c] "Energy Growth Rate to Slow Down," *China Economic Net,* December 17, 2004, http://en.ce.cn; Vandana Hari and Winnie Lee, "China's Oil Demand Jumped 15 Percent Last Year," Platts *Oilgram News* 83, no. 36 (February 23, 2005): 2, in quoting the official Xinhua News Agency, reported crude oil production in 2005 was predicted to ease to about 172 million metric tons (3.44 million b/d), down from 174.73 million metric tons (3.49 million b/d) in 2004. Surprisingly, the International Energy Agency (IEA) reports that production of crude oil during the first two quarters of 2005 surpassed these annual estimates by a considerable margin, averaging more than 3.6 million b/d, for a gain of 180,000 b/d over the first quarter of 2004. See IEA, *Monthly Oil Report,* August 11, 2005, table 3, p. 47.

National Petroleum Corporation (CNPC), China Petrochemical Corporation (Sinopec), and National Offshore Oil Corporation (CNOOC). Under this new structure, the companies have more geographic focus: CNPC in the north and west, and Sinopec in the south, and CNOOC in offshore exploration and production developments. Following their respective strengths prior to the reorganization, CNPC leans more toward upstream activities and Sinopec more toward downstream, especially refining.

CNPC is China's largest oil and gas company, accounting for 79 percent of oil supplied to the domestic market, 95 percent of the domestic natural gas market, and 40 percent of the market for petroleum

Table 2.4
Oil Consumption in China, Selected Years, 1980–2005

Year	Million b/d
1980	1.77
1990	2.23
1995	3.36
1999	4.36
2000	4.80
2001	4.92
2002	5.16
2003	5.55
2004	6.43[a]
2005	est. 6.75[a]

Source: Data for 1980–2003 taken from U.S. Department of Energy, Energy Information Administration (EIA), *International Energy Annual 2003*, table 1.2: World Petroleum Consumption 1980–Present, http://www.eia.doe.gov/pub/international/iealf/table12.xls.

[a] Estimates from International Energy Agency (IEA), *Oil Market Report*, August 11, 2005, table 2, "Summary of Global Oil Demand," p. 46. Chinese demand is defined by the IEA as the sum of domestic refinery output and net product imports, plus adjustments for direct crude burning, smuggling, and unreported refinery output. The Department of Energy's Energy Information Administration (EIA), in its *Short-Term Energy Outlook*, September 2005, table 3, presented higher estimates for oil consumption for 2004–2005: for 2004, 6.5 million b/d, and for 2005, 7.0 million b/d. The difference most likely reflects different methodologies, plus EIA access to more recent data. Additionally, both IEA and EIA have revised their estimates in their monthly reports reflecting new available data and potential effects of high prices.

products. Sinopec is the largest producer of refined oil products in China and in Asia. CNOOC accounts for more than 10 percent of China's domestic crude production. All three carried out initial public offerings between 2000 and 2002.

Over the years attempts to judge correctly future oil demand in China have met with only marginal success—a limited ability that continues today. This shortcoming was particularly evident in 2004 when actual demand turned out to be far in excess of what had been anticipated at the beginning of that year. This surprise, by pressuring available supplies, was among the key factors in pushing world oil prices to levels of $55 per barrel and higher. Clearly, it is the paucity of relevant data and the inaccuracy and limitations of the published data that preclude forecasting with any degree of confidence.

SECURITY OF SUPPLY: IS IT ENOUGH?

The dramatic increase in volumes of imported crude oil (see table 2.5) and in turn rising dependence on the sustainability of such imports over the past several years have heightened China's concern for supply security and have sent its oil companies on a worldwide search for equity oil. But security of supply, through diversity of supply, is not the only driver for China. Sustainability of oil flows is equally critical, as it is for both oil importer and oil exporter. Moreover, in a larger sense China seeks stability—economic stability, social stability, and, importantly, political stability.[7]

Supply diversity is important because it presumably offers protection against the loss of oil from a particular source or grouping of sources. It is, in its own way, an approach that seeks to provide insulation from the always volatile world oil market. Does supply diversity offer any guarantee against price volatility? Not at all. When oil prices rise or fall, for whatever the reason, they rise or fall everywhere. No importer, no exporter stands in isolation from the world oil market. Because of that, the goal of security of supply can be a false and misleading goal, if security of supply embraces just physical volumes and not prices paid for the oil consumed.

To illustrate, consumers were not faced with physical shortages of oil during 2004. There was no limit to purchases, say, of gasoline. Consumers could buy as much as they wished. Rather, any constraint or limitation to volumes purchased was expressed by price. Was the buyer willing and able to pay the asking price? In the United States, for example, growth in gasoline consumption was not deterred by a higher price, affirming that use of this commodity had effectively become price inelastic.

Although oil importers nonetheless continue to be preoccupied with security of supply, oil exporters in turn pursue their version of security by seeking diversity of markets. Thus, the loss of a particular market, should that happen, would not necessarily constrain sales and income. Beyond that, oil-exporting countries want to ensure the availability of demand for their oil before embarking on major investment projects to expand production capacity.[8] The world oil market today is constrained by refinery capacity limitations, particularly so in the United States where no new oil refinery has been built in 29

Table 2.5
China's Gross Imports of Crude Oil, Selected Years, 1990–2004

Year	Million b/d
1990	0.058
1993	0.330
1997	0.706
2003	1.664
2004	2.346

Sources: Figures for 1990, 1993, and 1997 from International Energy Agency (IEA), *China's Worldwide Quest for Energy Security* (Paris: OECD/IEA, 2000), p. 78; figures for 2003–2004 from IEA, *Oil Market Report,* February 10, 2005, http://www.iea.org, p.9.

years. Absent adequate refining capacity, oil exporters postulate, why then expand crude oil production?

Saudi Aramco, for example, exports about 500,000 b/d of oil to China and would like to sell more, but Saudi exports are limited by China's refining capacity. To assist China in expanding that capacity, Aramco, working together with ExxonMobil and Sinopec, plans to raise the current 80,000 b/d capacity of a refinery in Fujian province to about 240,000 b/d.[9] This expansion serves the national interests of both exporter and importer and, once again, emulates the original "oil for the lamps of China" approach taken by Standard oil in the early years of the twentieth century: if there isn't a market, create one.

Notes

[1] Unless otherwise indicated, the information presented in this background section was derived from Central Intelligence Agency (CIA), "China Oil Production Prospects," ER 77-10030U, June 1977. A somewhat different "take" on China had been given earlier in CIA, "The International Energy Situation: Outlook to 1985," ER 77-10240 U, April 1977. Therein it was noted that in China, the reserve and production outlook was much less favorable than it appeared a few years ago. Growing domestic needs, resulting from economic growth and trouble with coal production, would reduce oil exports to a negligible level by 1985, with exports in 1980 totaling no more than 500,000 b/d.

[2] CIA, "The World Oil Market in the Years Ahead," ER 79-10327U, August 1979, appendix C.

[3] Fereidun Fesharaki, "China's Downstream Petroleum Industry: Recent Developments and Future Potential," paper presented at the China Conference, CSIS, Georgetown University, Washington, D.C., April 3, 1986, p. 13.

[4] It is interesting to note that in February 1986 China had pledged to hold its oil exports to the 1985 level in support of OPEC. See Fesharaki, "China's Downstream Petroleum Industry," p. 26.

[5] Ibid., table 7, p. 28.

[6] International Energy Agency (IEA), *China's Worldwide Quest for Energy Security* (Paris: OECD/IEA, 2000), tables A-1 and A-2.

[7] "Chinese Economy: Stability First," *China Economic Net,* May 24, 2005, http://www.ce.cn/.

[8] Mohammed bin Dhaen al-Hamli, the United Arab Emirates (UAE) oil minister, reportedly said that the UAE planned to expand its producing capacity beyond 2.8 million b/d, but wanted guarantees of future demand, particularly from Asia, before making an investment in new producing capacity. See Kate Dourian, Vandana Hari, Alex Lawler, Shiva Lingam, "Producers Offer Qualified Support to Large Asian Buyers," Platts *Oilgram News* 83, no. 5 (January 7, 2005): 1.

[9] Simon Romero and Jad Mouawad, "Aramco Seeks India and China Ties," *International Herald Tribune,* February 18, 2005.

CHAPTER THREE

CHINA'S REFINING CAPACITY AND ITS OWN STRATEGIC PETROLEUM RESERVE

REFINING CAPACITY

For some years China enjoyed a refining capacity surplus that exceeded the country's needs. Then, as for so many other countries, that surplus disappeared. Now China must build new capacity aligned with its growing demand for petroleum products while keeping its dependence on imported petroleum products within acceptable limits.

China's capacity to refine crude oil is estimated to have been about 6.2 million b/d in 2004,[1] of which about 5.6 million b/d are operated by Sinopec and PetroChina. Small local plants account for about 10 percent of available capacity. Interestingly, the 6.2 million b/d capacity equates reasonably well with oil consumption that year. But, utilization of the key refineries is no more than 88 percent, which means that there is a gap to be filled through the importation of petroleum products. Key oil refineries are to increase their processing of crude oil by 9 percent in 2005, up 246,000 b/d over volumes handled in 2004, implying anticipated further growth in oil demand—and imports—during 2005. China refined 5.46 million b/d of crude oil in 2004, an increase of 13.7 percent over 2003, for the fastest growth rate in three decades.

As oil demand grows, so must the ability to refine increasing volumes of crude oil. If the refining sector fails to match demand growth, then the resultant gap in petroleum product supply will have to be covered by importing more products. A higher dependence on imported petroleum products translates into a higher national vulnerability. Petroleum products of the needed variety and specifications may not always be available from foreign suppliers. Disruption in crude oil supply can be offset by imports from other sources, or by withdrawals from strategic reserves, should the disruption be severe enough. Strategic reserves do not embrace petroleum products, however, and working product inventories are counted in days, not weeks.[2]

Perhaps with all this in mind, the CNPC has indicated that China's annual refining capacity will be expanded by 2 million b/d by 2010. Assuming little or no growth in domestic crude oil production, this expansion will have to be fed by imported crude oil.

The rapid expansion in oil demand and imports has, not surprisingly, attracted foreign investment to China. As noted, Aramco and ExxonMobil have entered into a $3.5 billion refinery and chemical joint venture with Sinopec in Fujian province, and Aramco is considering taking a position in a refinery being built by Sinopec in Shandong province.[3] The Fujian province refinery will be able to handle imported sour crude, whose share of world crude oil supplies is increasing.

Additionally, capacity at the Dushanzi refinery is to be expanded to enable processing of the volumes of crude oil to be delivered by the Kazakhstan-China pipeline.

STRATEGIC PETROLEUM RESERVE—CHINESE VERSION

China is responding to recommendations by the International Energy Agency (IEA) to establish its own strategic petroleum reserve as a buffer against supply disruptions. It is anticipated that the program will take in 110,000 b/d through 2008 and as much as 250,000 b/d by 2015, with the cost placed at $1.6 billion.[4] The oil, to be stored in above-ground steel tanks, is to provide a cushion of 20 to 30 days, based on refinery demand. The Chinese government currently requires commercial crude oil inventories to represent about 22 days of annual requirements, up from previous levels of around 15 days.[5] Still, national and commercial stocks combined fall well short of levels required for most industrialized countries.

China's growing demand for oil and the consequent increasing reliance on imports have outrun those estimates of supply and demand employed during the development of the country's initial petroleum reserve strategy. At the end of the 1990s, China expected annual petroleum imports to average 2 million b/d by 2010. On that basis, plans were formulated to establish a 25-million-ton oil reserve that would be the equivalent of 90 days of oil imports.[6] As is known, crude oil imports in 2004 of 2.45 million b/d surpassed the level anticipated for 2010 by almost 25 percent. To provide a cover today equal to 25 days, based on 2004 crude oil import levels, the reserve then would have to be roughly 25 percent higher than the volume anticipated for 2010.

The strategic reserve will be located next to China's four largest oil refining centers and will take heavy, sour Middle East crudes plus lighter West African crudes. Work on the reserve began in 2005, with storage expected to begin by the end of the year.[7] Filling a tank farm adjacent to the Zhenhai refinery, where construction began last year, will be the first step in establishing the reserve, with the fill to reach 22 million barrels by October 2006.[8] Construction has started on three other sites, where a capacity to hold 100 million barrels is to be ready by the end of 2008.

Whereas the Strategic Petroleum Reserve of the United States and those of other member-countries of the Organization for Economic Cooperation and Development generally are to be tapped only in the event of a serious supply disruption, China has indicated that it will turn to its stockpile when and if necessary to mitigate the impact of rising oil prices on the profitability of state-owned companies. Having said that, China does intend to look to its reserve to cover any shortfalls deriving from disruptions in oil import levels.

NOTES

[1] Reuters, "China's Top Refineries Expected to Raise Output to Meet Strong Demand Growth," Singapore, February 8, 2005.

[2] The CNPC has indicated that by 2010 annual refining capacity is likely to increase by 30 percent or 86.3 million metric tons (MMt), to 375.3 MMt, equivalent to 7.506 million b/d. See Angelina Lee, "TNK-BP to Ship Only Crude, Not Products, to China," Platts *Oilgram News* 83, no. 43 (March 4, 2005): 2. That increment of 86.3 MMt or 1.73 million b/d implies 2004 refining capacity of about 5.8 million b/d.

[3] "Aramco May Buy into Sinopec Site," Bloomberg, London, January 14, 2005.

[4] UPI, "China Plans Oil Reserve to Manage Prices," Beijing, February 2, 2005.

[5] Winnie Lee, "China Sets Construction for Strategic Oil Terminal," Platts *Oilgram News* 83, no. 49 (March 14, 2005): 3.

[6] "Adjusting Petroleum Reserve Strategy," *China Economic Net,* May 26, 2005, quoting the publication *Economic Daily*.

[7] Xinhua News Agency, "China to Fill Strategic Oil Reserve," July 1, 2005, http://news.xinhuanet.com/english/2005-07/01/content_3170702.htm.

[8] "Strategic Oil Reserve to Moderate Prices," *Shenzhen Daily,* February 2, 2005.

CHAPTER FOUR

CHINA'S OIL SECTOR: YESTERDAY, TODAY, AND TOMORROW

THE OIL SECTOR IN 2004

The sharp and unexpected growth in oil imports by China during 2004 has been taken by most observers as one of the key factors behind the dramatic oil price rise during the year. China disagrees with this negative implication and instead underscores that such imports reflect growing economic globalization and benefit both the nation and exporters. The minister of the National Development and Reform Commission has noted that "by importing raw material and energy we contribute to resource-rich countries' economic growth."[1] In support, it has been noted that for China its share of world oil growth in demand in 2004 did not differ substantially from the share held during the preceding decade.

Although most media coverage highlights the growing dependency of China on crude oil imports, China has been a trader in both crude oil and petroleum products—that is, it has been both a buyer and a seller. Trade in petroleum products plays a comparatively important role in the country's oil balance, but of course much less so for crude oil.

Table 4.1 offers a simplified approach to defining oil supply and demand in China, employing, wherever possible and appropriate, data released by Chinese governmental organizations. Official data on the Chinese oil and energy sectors are quite limited in scope and even then are often of questionable accuracy. Published reports by Chinese research institutes and universities are useful and sometimes more reliable. Nonetheless, the energy sector as a whole lacks the transparency needed for confident evaluations.

China has been exporting crude oil to a variety of countries, chiefly Japan, for a number of years, mainly in keeping with long-term agreements. However, in view of growing crude oil imports, Chinese officials questioned the rationality of continuing these sales. Consequently, crude oil exports are to be reduced by two-thirds during 2005.

Petroleum product trade remains substantial, with net product imports during 2004 averaging 528,400 b/d. Net product imports thus represented roughly 9 percent of petroleum products consumed in China during 2004, a relative dependency equating to that of the United States in terms of product imports—also 9 percent—in 2004.[2] Exports of petroleum products are mainly for the oil-processing trade, so it has been explained.[3]

The approach illustrated in table 4.1 defines the volume of petroleum products available for domestic consumption, but excludes the "losses, etc." in the crude oil category and fails to account for refinery losses and the like—line items that should be counted in any consumption estimate.[4] Their omission in effect understates the amount of oil the economy uses.

A more simplistic approach to calculating 2004 oil consumption, again using table 4.1 data, would seem to be in order:

- Production of crude oil, plus net crude oil imports: 5.83 million b/d,
- Plus net petroleum product imports: 0.53 million b/d,
- Resulting in an average of 6.36 million b/d available for domestic use during 2004.

The derived 6.36 million b/d, in rounded terms, equates with the 6.4 million b/d of oil demand in China as calculated by the IEA for the year 2004.

As noted, care must be exercised when working with Chinese data. When Chinese sources refer to the consumption of "refined petroleum products," Western researchers would assume that this reference could literally be taken as total annual oil product consumption. Not so. For example, it was reported that refined petroleum product consumption estimated for 2005 would be 3.4 million b/d, up by 8 percent to 9 percent over the 2004 level, which in turn had risen by 19 percent over the 2003 level.[5] Then, fortunately, the text subsequently

Table 4.1
Estimated Oil Supply and Demand in China, 2004

	Million metric tons	Million b/d
Crude oil supply		
Production	174.50	3.49
Imports	122.71	2.45
Exports	5.49	0.11
Losses, etc.	18.64[a]	0.37
Subtotal	273.08	5.46
Petroleum product supply		
Charge to refining	273.08	5.46
Refinery yield	251.23[b]	5.02
Product imports	37.88	0.76
Product exports	11.46	0.23
Available for domestic use	277.65	5.55

[a] Losses in the field, burned directly as a fuel, and added to storage. Derived as the difference between reported crude oil production and reported charge to refining.

[b] Estimated as equal to 92 percent of refinery charge.

Sources: "China's Import of Crude Oil in 2004 Hits 122.72 Million Tons," *China View*, http://www.chinaview.cn, January 25, 2005; Winnie Lee, "Chinese Crude Imports to Grow 11 Percent This Year: CNPC," Platts *Oilgram News* 83, no.15 (January 24, 2005); AFX Press, "China's 2004 Diesel Output up 19.5 pct, Gasoline up 10.2 pct – Report," http://www.afxpress.com, January 25, 2005; www.ce.cn, January 25, 2005; and *Oil and Gas News*, "China Processes Record Crude," http://www.oilandgasnewsonline.com, January 17–23, 2005.

defined "refined petroleum products" as gasoline, diesel, and kerosene, but thus omitting consumption of fuel oil and lubricants, among others.

Trying to understand total Chinese oil demand, as reported by the official Xinhua News Agency, is much more frustrating.[6] Under most circumstances, "net oil supply" would be those volumes available for domestic use. Net oil supply, as reported by Xinhua, represents something different. In other words, the total includes crude oil produced, plus volumes imported minus those volumes exported, *plus* products derived from refining the net volumes of crude oil, plus imports of petroleum products, minus product exports. Thus, for 2004 Xinhua reported net oil supply to be 507.05 million tons, or the equivalent of 10.14 million b/d, far in excess of actual oil consumption. The casual observer or researcher, not familiar with China's true oil supply and demand balance, would be seriously mistaken if this version of net oil supply were accepted without further investigation.

WHAT ABOUT 2005?

China, facing rising oil imports, high coal prices, and continued electric power shortages, has established a high-level energy coordination panel, led by Premier Wen Jiabao. This task force, which is not regarded by officials as a precursor to a cabinet-level energy ministry, is to take the lead in developing a responsive energy policy that, among other issues, will work to[7]

- Secure overseas oil and gas reserves,
- Ease the chronic electricity crunch,
- Stabilize the supply of coal,
- Enforce industrial energy efficiency, and
- Promote nuclear power and other renewables.

Above all, energy security is the key task.

Apparently China is under great pressure to moderate growth, centered in part on saving energy, while eliminating current energy bottlenecks, especially electric power shortages, reducing energy waste, and correcting low efficiency in use.

Importantly, China wants to control its reliance on oil imports, currently on the order of 40 percent of supply, and to reduce that reliance to 35 percent and hold it there.[8] A portion of the decline would come about because high world oil prices would bring investment and advanced technology to China's oil sector.

The head of the National Development and Reform Commission saw demand side management as important as, for example, increasing electricity supply while providing for greater coal mine safety.[9] China also has emphasized that construction timetables of key energy projects will be advanced, including coal liquefaction, oil and natural gas development, gas-fired power plants, and renewable energy.[10]

There had been an emerging consensus at the beginning of 2005 that, although China's growth in demand for oil that year, to be satisfied basically through imports, will decline somewhat in relative terms, the absolute increment will still be considerable, possibly on the order of 300,000 b/d or more (see table 2.4).

CNPC anticipated crude oil imports averaging 2.8 million b/d during 2005, a gain of 11 percent over 2004.[11] The market also would want some appreciation of what total oil consumption might be during 2005, and that is not easy to come by. Again, the difficulty derives in the way that China presents data relating to the oil sector. For example, CNPC emphasizes, as do other Chinese government sources, demand for crude oil, not total product consumption.

Nonetheless, the data most relevant in today's world oil market are the projected growth in imports of crude oil. Most media reporting on Chinese oil dwell on those volumes of crude oil imported and ignore net trade in both crude and products. Omitting data on exports of crude oil and net trade in petroleum products understates the dependence of China on supplies of foreign oil and understates its "take" from the world oil market.

At the beginning of 2005:

- Crude oil demand in 2005 was placed at more than 6.4 million barrels a day.
- But that would include both domestic production and imports.
- If crude oil imports averaged 2.8 million b/d (see table 4.1),[12] then domestic production should average 3.6 million b/d for the year—that is, the difference between crude oil demand and crude oil imports, implying a gain of some 100,000 b/d over 2004.
- Demand for petroleum products is placed by CNPC at about 3.4 million b/d, a gain of 260,000 b/d over 2004.
- Clearly, that is not petroleum product demand as normally defined. As noted above, reference most likely is made just to the demand for gasoline, diesel, and kerosene, and excludes other products.
- Unless data are provided on net trade in crude oil (relatively small and declining) and on net trade in petroleum products (quite substantial), then reasonably confident oil consumption levels cannot be determined.

A senior official of CNPC noted that the consumption of crude oil would jump to 6.4 million b/d in 2005, a gain of almost 12 percent over the 5.76 million b/d consumed in 2004.[13] Given that piece of data, if net trade in petroleum products could be reasonably estimated, then total oil consumption could be derived, employing the simplified methodology described above.

There was some good news to report relative to crude oil production during the first quarter of 2005. Domestic crude oil production was up 180,000 b/d compared with first quarter 2004, averaging nearly 3.67 million b/d. Oil fields in western China and offshore provided the bulk of this gain. If this gain can be preserved throughout the year as a whole, requirements for imported oil will be comparably reduced, assuming no upward revision in oil demand.[14]

The IEA has projected Chinese oil demand in 2005 to average 6.75 million b/d, down from an initial estimate of 6.85 million b/d. Available evidence supported that projection, implying an increment in demand that year of roughly 350,000 b/d. That amount is less than the previous year, but still high enough to support maintenance of 2004 oil prices throughout 2005, especially if additions to world oil supply from countries not members of the Organization of Petroleum Exporting Countries (OPEC) are less than 2004 additions, which they likely will be.[15]

Slackening in the growth in Chinese oil demand could be attributed to the coming online of new electric power generating capacity and to the closure of a number of less efficient industrial plants. At the same time, more specific measures, also dedicated to conservation, have been introduced. For example, the city of Beijing for the first time has introduced peak-hour charges for electricity that are more than quadruple charges levied during low-use periods.[16] In addition, to avoid market speculation and hoarding of motor fuel, Sinopec is offering to consumers coupons that would be bought at the current market price, to be used to purchase fuel at a future date, within a fixed timeframe of three months.

China's installed electric power generating capacity and annual output is about one-half that of the United States. Nonetheless, installed capacity has increased rapidly over the years (see table 4.2).[17]

Table 4.2
China's Electric Power Generating Capacity, Selected Years, 1987–2004

Year	Capacity (million kW)
1987	100
1995	200
2004	440
2005 (est.)	500–510

As the first four months of 2005 came to a close, available data did not yet provide a sufficiently clear picture of the year as a whole.[18] Imports of crude oil rose 35 percent, compared with the first four months of 2004, to 2.55 million b/d.[19] At the same time exports of crude oil rose almost 35 percent to 150,000 b/d. Product imports declined as product exports were stimulated by the cap on domestic prices.

Conjecture had held that China's refineries were processing more crude to meet domestic demand and were reducing imports of refined products. Demand growth had not been particularly strong, permitting crude oil exports to jump, taking advantage of strong overseas margins.

The import/export pattern was reversed in May 2005, as the government took steps to halt exports of diesel fuel and to cut gasoline exports in half.[20] This shift in import/export patterns further demonstrates the inadvisability of extrapolations based on a very short time period and for a country where government decisions, not market decisions, carry the day.

China's domestic oil market may be quite different in the coming years. China plans to fully open its petroleum and refined oil market to foreign competition by the end of 2006 in response to commitments made when it entered the World Trade Organization.[21] What will that mean for the prices on the domestic market? The country's oil prices are still controlled by the government, but with partial adjustment according to world oil market prices. Indeed, China in late May 2005 reduced gasoline prices in keeping with fluctuations on the world market. Currently the price for domestically refined crude oil is about $16.44 per barrel less than the international price. Nonetheless, the end of 2006 is still a bit too far away to make any authoritative oil price predictions.

In 2004, the broad and continuing shortage of electricity led many consumers to turn to backup powers generators that burn diesel and fuel oil. The IEA thought that perhaps on the order of 250,000 to 300,000 b/d had been consumed during the year by such backup generators.[22]

China had been expected to experience some electricity shortages again during 2005, which implied that consumers would turn to reliance on small diesel generators to help fill the supply gap as they have in the past. Partly in recognition of this trend, in March 2005 the IEA revised upward its estimate of oil consumption growth in China during 2005 to 500,000 b/d, representing an additional requirement of 100,000 b/d.[23] That revision in turn raises oil consumption to 6.88 million b/d. In April 2005 the IEA pointed to slowing growth in Chinese oil demand, among other developments, that should help lower prices.[24]

In summer 2005, electric power shortages did hit, causing rolling blackouts, but the shortages were less severe than originally expected. This is due in part to a cooler, wetter summer, which in turn provided greater hydropower, more coal-fired plants online, fuel switching, and government-enforced industrial slowdowns. The China Electricity Council estimated a shortage of 25 million kW in the third quarter.[25]

Reflecting data for the year's first six months, the Energy Information Administration (EIA) revised its forecast for Chinese oil demand during 2005 to 7.0 million b/d.[26] That revision in turn places the EIA estimate of growth in China's oil use in 2005 at 500,000 b/d.

OPEC also revised downward its projection of China's daily oil consumption to 6.85 million b/d, reflecting new assumptions about lower 2005 demand growth. The OPEC report reiterates the difficulty of assessing demand from developing Asia caused by timeliness and quality of data.[27]

Trying to find common ground between the IEA, EIA, and OPEC forecasts of future Chinese oil demand and oil import requirements is a thankless task. Researchers perhaps would be better served by selecting either the IEA or the EIA as the primary source for both current data and for forecasts. Even so, a comparison of IEA and EIA forecasts out to 2025 and beyond yields some common ground.

If the Chinese demand for oil is to be satisfied, OPEC seemingly has accepted that it will have to utilize close to its maximum producing capacity.[28] Moreover, OPEC also recognizes that Russian oil production is slowing, which in turn may mean that expansion of its exportable oil surpluses will also slow, in relative terms, further enlarging the call on OPEC oil.

Meanwhile, China is closely watching the tenuous relationship between Iran and the United States, the increasing international concern over the inability to secure a lasting peace in Sudan, and the cooling climate between the United States and Venezuela. Further deterioration could well affect not only oil availability but also political relations between China and the United States.

Iran is a major supplier of oil to China, having accounted for 13.6 percent of oil imports in 2004.[29] Sudan is comparatively less important as an oil supplier to China, having provided about 6 percent of China's imports in 2004. Nonetheless, the prospect remains that any further, serious deterioration in the situation inside Sudan or between the United States and Iran might lead to the question of raising sanctions against either or both at the United Nations. China has a seat on the UN Security Council, where voting on actions to be taken, or rejected, has to be unanimous, and would thus find itself in an unwelcome position. Nonetheless, given the continuing tenuous balance between world oil supply and demand, a UN-backed oil-related sanction imposed on either Iran or Sudan would have to be considered unlikely.

FORECASTS FOR 2010

Observing China's rising requirements for imported oil, and the high prices paid for these imports, leads to a simplistic conclusion: China is the victim of its own successes. Not only has it had to pay higher prices for imported oil, these high prices have caused China to pay high prices for equity oil positions around the world. The growth in China's GDP and the public's newly acquired capabilities to enjoy a much higher lifestyle, including automobile ownership (there are currently 24 million cars in China), have pushed oil demand upward. Because domestic oil supply has been relatively static, reliance on foreign oil is rising.

Not all the blame for high prices should be laid at the doorstep of China. Had OPEC member-countries not gradually worked off their spare producing capacity during the past decade or so, to the point

Table 4.3
China's Imported Oil: Transportation Diversification

Means of Delivery	Percent of Total 2010	Current[a]
Ocean-going tanker	83.0	93.5
Pipelines	15.0	0.0
Rail	2.0	6.5
Total	100.0	100.0

[a] Implied.

where the ability to respond quickly to higher world oil demand had essentially disappeared, then the 2004 surprise of the measure of China's need for foreign oil might have been covered just by bringing that spare capacity into play, presuming of course the willingness to do so. Absent spare producing capacity, pressure on available supplies increased and prices rose accordingly.

Thus, rising Chinese oil demand and the need for more and more foreign oil to fill the gap, coupled with minimal spare producing capacity worldwide, will influence prices throughout 2005 as in 2004.

China's continuing efforts to expand the number of foreign sources of oil supply as well as diversify how oil imports find their way to China are expected to show some successes by 2010. Table 4.3 shows the transportation diversification for China's oil distribution as noted by a senior PetroChina official.[30]

Developing pipelines to deliver crude oil and natural gas to China during this decade is perhaps the most observable success. This availability will reduce reliance both on ocean-going tankers and on imports by rail and will help give China the diversity that it seeks. Completion of the oil pipeline now being laid from western Kazakhstan and the construction of the planned Taishet-Skovorodino pipeline to support oil imports from Russia will make these shifts possible.

Elsewhere it has been noted that growing oil imports partly reflect China's inability to find and develop new sources of domestic production sufficient in size not only to offset declines at producing fields but to provide for growth in national production. There are just two major oil fields in China: Daqing, producing 976,000 b/d (see table 4.4) but declining; and Shengli, where production averaged 534,800 b/d in 2004, having held stable for the past several years.[31]

Table 4.4
The Decline and Fall of Daqing

Year	Million b/d
2003	0.976
2005 est.	0.908
2010	0.600
2020	0.400

Sources: Figures for 2003 and 2005 from AFX Press, "China's 2004 Diesel Output up 19.5 pct, Gasoline up 10.2pct—Report," http://www.afxpress.com, January 25, 2005; Maurizio D'Orlando, "Oil Imports Increase Again," http://www.asianews.it, January 15, 2005; "Shengli Oilfield Produces 27 mln Tons of Crude Oil in 2004," *China Views*, http://www.chinaviews.cn, January 11, 2005. Figures for 2010 and 2020 from James Kynge, "The Daqing Donkeys that Give the Nod to Russian Oil," *Financial Times*, August 23, 2004.

Daqing has been the centerpiece of Chinese oil production since its discovery, but now that field is in a steady decline that is not reversible. Whatever growth can be secured from other fields currently is barely sufficient to offset Daqing losses. Holding national production constant is probably the best that can be hoped for, until major new oil finds can be made and brought into play.

In March 2004, Chinese state media reported that Daqing planned to cut output by 7 percent a year between then and 2010, reflecting concerns that its proven reserves were facing exhaustion.[32] That projected decline must have been unacceptable to the bureaucracy, for several months later it was announced that these planned production cutbacks would be postponed by finding new reserves and boosting natural gas output.

Where will future oil production be found in China? In the past East China (primarily Daqing) has led the way, followed by western China and offshore fields. But, with the decline of Daqing, among others, the relative contribution of East China has given way to western China and particularly to offshore fields, where the largest relative and absolute gains have been recorded (see table 4.5).

Initial success has been found in the west, in the Xinjiang Uygur Autonomous Region, where crude oil production will likely reach 1 million b/d by 2010, up from 440,000 b/d in 2004.[33] This gain, if realized, will offset declines at Daqing and elsewhere and should provide the basis for some growth of the national total.

Table 4.5
Regional Shifts in China's Crude Oil Production, 1990 and 2003

	1990		2003	
	Production[a]	Percent of Total	Production[a]	Percent of Total
East China	2.53	91.6	2.21	65.3
West China	0.21	7.5	0.73	21.7
Offshore	0.03	0.9	0.44	13.0
Total	2.76	100.0	3.38	100.0

Source: Xinhuanet, "Offshore, Western Oil to Sustain Country's Output Growth," http://news.xinhuanet.com/english, May 25, 2005.

[a] Million barrels per day. Because of rounding, data may not always agree with production levels noted elsewhere in the text.

A senior Chinese oil expert has noted that by 2010 the production of oil from offshore fields and from fields located in the western regions may account for more than one-half of the country's total oil output.[34] Indeed, he added, these oil fields will allow a stable growth in Chinese oil production for the next 5 to 15 years. Nonetheless, production will hit its peak at some point during 2010 to 2020. That peak likely will be in the 3.4 million to 3.6 million b/d range and could hold beyond 2020 for some 10 years, if major new discoveries are made and if breakthroughs occur in the development of unconventional oil.

Prospects for any substantial future growth in oil production beyond 2010 are relatively limited. The China Economic Information Network, under the State Information Center, projected domestic oil production peaking at 4 million b/d around the year 2015.[35] That of course translates into even higher demand for foreign oil beyond then, assuming continued expansion in economic activity.

A larger question remains. Will economic growth and in turn growth in demand for imported oil continue? Yes, but not at past demonstrated levels. A very limited view of the future—in this instance, the year 2010 is the future—has been offered by Chinese authorities.[36]

- Oil demand in the year 2010 will fall within the range of 7.0–7.6 million b/d.
- Oil imports in the year 2010 will fall within the range of 3.6–4.0 million b/d.

- These amounts imply that domestic oil production that year will fall within a 3.4–3.6 million b/d range.
- Natural gas imports will fall with a range of 20–25 billion cubic meters (bcm), up from zero in 2000.

Nonetheless, these estimates, at least for oil, already appear out-of-date. If oil demand, as defined by China, equates just to crude oil, then, given the dependence on petroleum product imports, oil consumption—or oil demand as referenced above—is understated.

FORECASTS TO 2025 AND BEYOND

There can be no doubt that China is very much concerned about its oil and energy future. The general manager of CNPC maintains that China's petroleum security will be in a grim situation at present and beyond.[37] What are the key concerns?

- The gap between oil supply and demand is widening, emerging as one of the constraints to the country's economic growth.
- Natural gas production will increase by nearly 10 percent, and the share of natural gas in energy consumption will increase by 10 percent (implying rising dependence on imports).
- Consumption of petroleum and petrochemicals will grow by a large margin. Local shortages occasionally occur.
- China has considerable room for energy conservation, but technologies are lacking and efficiency of energy use is a fraction of that of developed countries.
- A great geographic disparity between where oil is produced and where oil is consumed is growing. This is especially true for the Asia-Pacific region, where oil production accounts for 10 percent of the world total, but consumption reaches 30 percent.

China now stands second only to the United States in terms of daily consumption of petroleum products, having surpassed Japan in 2003. China finds it exceedingly difficult to correctly measure economic activity in a country of 1.3 billion inhabitants. How then to judge the future, a future that to a considerable degree will likely shape, and be shaped by, energy and oil supply and pricing?

Perhaps the easiest approach is to follow closely what others think, such as the IEA in Paris or the EIA of the U.S. Department of Energy. Both the IEA and the EIA have looked further into the future, out to 2025 and even beyond to 2030.

The latter organization, for example, projects China's oil demand to reach 12.8 million b/d by 2025, with 9.4 million b/d of that demand having to be met by imports.[38]

The IEA is a bit more constrained in its forecasts of future oil demand in China. For example, it forecasts demand to average 7.9 mmb/d in 2010, whereas the EIA anticipates demand reaching 7.2 million b/d by 2005. Further, the IEA forecasts 2020 demand at 10.6 million b/d and 13.3 million b/d in 2030. The midpoint of the IEA forecasts for 2020 and 2030 would be about 12 million b/d, which compares reasonably well, given all the limitations of long-term forecasting, with the 12.8 million b/d forecast by EIA for 2025.

China has its own view of crude oil and natural gas supply and demand in the year 2020. Crude oil production is placed at 4 million b/d, the maximum future domestic production. Demand is placed at 9 million b/d, implying an oil import requirement of 5 million b/d. That import requirement appears to be remarkably low, and the reader is referred to the previous discussion regarding Chinese definitions of oil supply and demand. Demand for natural gas in 2020 is placed at 200 bcm. An estimate of domestic production was not provided, but imports were projected to provide about 50 percent of supply. Even so, a gap of 8 bcm between demand and supply was anticipated.[39]

How can that future demand level be satisfied, given the limited prospect for meaningful increases in domestic crude oil production? Two options are available to the government—demand management and expanded reliance on imported oil. Although efficiency and conservation are being emphasized, efforts to develop equity oil outside the country show no signs of slowing. A measurable slowdown in economic growth, for whatever the reason, resulting in reduced oil demand, cannot and should not be ruled out.

Notes

[1] Song Quan, "Oil Imports Benefit Nation and the World," *China Daily*, March 8, 2005.

[2] The United States imported an average of 2.861 million b/d of petroleum products during 2004, but these imports were offset in considerable part by product exports averaging about 1 million b/d. Thus, on a net basis, the reliance of the United States on imported petroleum products averaged 9 percent during the year. See Department of Energy, Energy Information Administration (EIA), *Monthly Energy Review*, February 2005, table 3.3h, p. 55.

[3] "Crude Oil, Refined Oil Exports Decrease," *China View*, http://www.Chinaview.cn, February 17, 2005.

[4] In other words, it has been noted that product yields from crude oil charged to refining averages 94 percent. What happens to the unaccounted for 6 percent?

[5] "Refining Capacity Rise Lags Demand Growth," *China Economic Net*, December 22, 2004.

[6] Vandana Harti and Winnie Lee, "China's Oil Demand Jumped 15 Percent Last Year," Platts *Oilgram News* 83, no. 36 (February 23, 2005): 2.

[7] Reuters, "China to Form Task Force to Tackle Energy Policy," Beijing, March 7, 2005.

[8] "China to Control Its Reliance on Oil Imports," *China Daily*, http://www2.chinadaily.com.cn/English, April 23, 2005.

[9] "China to Set Up Energy Task Force as Planner Warns of Power Shortage," *China News Asia*, http://www.channelnewsasia.com, March 7, 2005.

[10] Reuters, "China to Form Task Force to Tackle Energy Policy," Beijing, March 7, 2005.

[11] Winnie Lee, "Chinese Crude Imports to Grow 11 Percent This Year: CNPC," Platts *Oilgram News* 83, no. 15 (January 24, 2005): 2.

[12] Crude oil imports averaged 2.407 million b/d during the first four months of 2005, a jump of 4.4 percent over first quarter 2004 crude oil imports. Conversely, petroleum product exports during that time period were down 14.9 percent, to 875,000 b/d. See Winnie Lee, "China's Imports Hit Record," Platts *Oilgram News* 83, no. 98 (May 23, 2005): 2, quoting preliminary trade data from the General Administration of Customs. For an indicator of the difficulties associated with working with Chinese data, another source stated that first quarter 2005 crude oil imports fell 1.6 percent from a year earlier, to 2.4 million b/d, again quoting customs data. See "Crude Oil Refining Rates Edge Up," *China Economic Net*, http://en.ce.cn/Industries/Energy&Mining, May 22, 2005. That level of imports is in line with the 2.4 million b/d average recorded in 2004. The inclination is to accept the latter data and to question whether the *Oilgram* report might have been based on an incorrect translation from the Chinese.

[13] Yahoo News, "Sharp Rise in Oil Demand Expected in Energy-hungry China," http://uk.news.yahoo.com/, January 21, 2005.

[14] See "China Saw Crude Oil Output Grow in 1st Quarter," *China Economic Net,* http://en.ce.cn/Energy&Mining, May 22, 2005.

[15] OPEC membership includes: Algeria, Indonesia, Iran, Iraq, Kuwait, Libya, Nigeria, Qatar, Saudi Arabia, United Arab Emirates, and Venezuela.

[16] Peak hours are between 11 a.m. and 1 p.m. and between 8 p.m. to 9 p.m., July to September.

[17] Asia Pulse/XIC, "China's Electric Power Sector Reaches Growth Limit," *Asia Times,* http://www.atimes.com, May 5, 2005.

[18] "China's Oil Import Bill Surges 86 Percent on Record Prices," *Bloomberg,* May 26, 2005.

[19] Imports from the Middle East led the way during the first quarter of 2005, followed by West Africa. Saudi Arabia was the leading supplier, averaging some 412,000 b/d in deliveries.

[20] "Fuel Exports Cut as Demand Picks Up," *China Economic Net,* http://en.ce.cn/Industries/Energy&Mining, June 3, 2005.

[21] Xinhua News Agency, "China to Open Petroleum Market by End of 2006," http://news.xinhuanet.com/english, May 26, 2005.

[22] Jeffrey Logan, "Energy Outlook for China: Focus on Oil and Gas," testimony before the U.S. Senate Committee on Energy and Natural Resources, February 3, 2005, p. 3.

[23] Alejandro Barbajosa, "Oil Demand Is Rising Faster than Expected, IEA Says," *Bloomberg,* March 11, 2005.

[24] The IEA publishes a *Monthly Oil Report*, thus offering the opportunity to revise its estimates 12 times yearly, if conditions warrant. That opportunity is put to regular use, reflecting the oil market lack of transparency and the absence of timely, precise data.

[25] Xinhua News Agency, "China's Electricity Shortage to Hit 25 Million Kilowatts This Year," *People's Daily Online,* http://english.people.com.cn, July 29, 2005.

[26] EIA, *Short-Term Energy Outlook,* August 2005, p. 2, and table 3, p. 14.

[27] Organization of Petroleum Exporting Countries (OPEC), *Monthly Oil Market Report,* August 2005, table 6, p. 24.

[28] Javier Blas and Kevin Morrison, "Chinese Demand Set to Push OPEC to Limit," *Financial Times,* February 17, 2005.

[29] Antoaneta Bezlova, "China-Iran Tango Threatens US Leverage," *Asia Times,* http://www.atimes.com, November 30, 2004.

[30] "Beijing: 15 Percent of Oil Imports to Come Overland by 2010," *Moscow Times,* April 21, 2005.

[31] "China's Second Largest Oil Field Maintains Stable Production," *China View,* http://www.chinaview.cn, March 6, 2005.

[32] "China to Halt Oil Output Cut," *Business Daily Financial Mail,* http://www.bdfm.co.za, July 19, 2004.

[33] "Xinjiang to be No. 1 Oil Production Area," *China Economic Net,* http://en-1.ce.cn/Industries/Energy&Mining, May 22, 2005.

[34] Xinhua News Agency, "Offshore, Western Oil to Sustain Country's Output Growth," http://news.xinhuanet.com/english, May 25, 2005.

[35] Xinhua News Agency, "China's Crude Oil Output to Peak at 200 mm Tons around 2015," *Alexander's Gas & Oil Connections* 9, issue 23 (November 25, 2004).

[36] *Business Report 2005,* "China Will Need to Boost Oil Imports to Meet Future Demand," February 16, 2005, quoting a director of the Energy Research Institute, Development and Reform Commission. It is likely that the referenced "oil" imports relate just to the import of crude oil. Another source projected oil consumption in 2010 to reach 7.6 million b/d. (See "Internationalizing China's Energy Industry," *China Economic Net,* January 31, 2005.) This source quoted the *Beijing News.* It is probable that all three sources noted herein based their reporting on the same original Chinese source.

[37] He Zhenhong, "Three Oil Magnates' Opinions on Energy," *China Economic Net,* http://en.ce.cn/Insight, June 1, 2005.

[38] See International Energy Agency (IEA), *World Energy Outlook 2004* (Paris: OECD/IEA, 2004), table 3.1.

[39] See Elaine Kurtenbach, "China Facing a Coal Shortage," *Business Week,* http://businessweek.com/ap/financialnews, May 25, 2005.

CHAPTER FIVE

ADDRESSING OIL IMPORT DEPENDENCY

DEMAND MANAGEMENT VERSUS IMPORT DEPENDENCY

China's commitment to demand management, including energy conservation, should not be taken lightly. A strategic plan for energy conservation—the China Medium and Long Term Energy Conservation Plan—is divided into two phases. The first phase covers the years 2006 to 2010; the second phase covers the following decade. The plan, drawn up by the National Development and Reform Commission, calls for holding total energy consumption below 3 billion tons of coal equivalent by the year 2020.

Public awareness, incentive policies, higher taxes, higher prices, and greater emphasis on renewable forms of energy are all to support the energy conservation effort. For example, renewable energy is expected to account for 10 percent of total energy consumption by 2010, compared with the current level of around 1 percent. Absent the effort to enhance energy security, however, energy demand is expected to exceed 4 billion tons of coal equivalent that year. In other words, an annual energy conservation of 3 percent between 2003 and 2020 is anticipated.[1]

China is extremely wasteful in terms of units of energy consumed to secure growth in its GDP. It is the combined impact of poor regulation, limited skills, and inferior technology that lead to squandering of available resources and in some instances to physical shortages.[2] For example, to produce one ton of steel, China requires double the energy of Japan and Korea. The average national recovery rate for coal stands as low as 30 percent and for some small-scale private mines only 15 percent. At the same time, the recovery rate, as established by the government, should be more than 75 percent. In sum, for every dollar of GDP, China reportedly consumes three times as much energy as the global average.[3]

The U.S. experience with demand-side management has not gone unnoticed in China. Chinese energy experts and policy professionals have been sent to California to study demand side management ideas of that state.[4] Additionally, a China-U.S. Energy Efficiency Alliance, made up of government, corporate, and individual participants has been established to address the critical energy challenges that China now faces.

HEADLINES TELL THE STORY FOR CHINA

The scope and depth of China's program to develop substantial equity oil positions around the world are amply demonstrated by three headlines of the day:

- "Chinese oil firm bids for Unocal" (*Los Angeles Times,* June 23, 2005)
- "Putin in China to talk oil and weapons" (*AsiaNews,* October 14, 2004)
- "China's oil sands role tests U.S." (*Globe and Mail,* December 31, 2004)

Then there are other headlines, reflecting the concerns of a competing oil importer whose future in large part also rests on gaining access to needed supplies. These headlines read: "India finds itself trailing in fight for global energy deals."[5] "China is ahead of us in planning for its energy security—India can no longer be complacent," said Indian Prime Minister Singh during a recent speech.[6]

The China-India competition has led some observers to speak of an "Axis of Oil," a play on the "Axis of Evil" referenced by President Bush in his State of the Union message given in January 2002.[7] This "Axis of Oil" would embrace China, India, and Russia and would involve expanding energy and military trade, with closer political linkages following.

It is generally accepted that private sector competition for markets is a good thing, just as is cooperation between governments. Competition between governments—for example, between China and India for access to oil supplies—disadvantages private sector companies, especially where state oil companies are involved. Private sector cooperation that distorts the markets is no more a good thing than government competition.

Both China and India understand that competition between the two countries for access to oil supplies does not serve their best national interests. Consequently, the two countries will instead work toward cooperation that may secure cheaper prices from producers and rights to overseas oil assets.[8] It would be surprising, though, if much cooperation were to emerge from this agreement, considering that national interests will always dominate.

India mirrors China in many ways. Crude oil production is stagnant, at around 640,000 b/d, and imports fill the expanding gap between domestic supply and demand. Today, Indian oil import reliance stands at 70 percent and may hit 85 percent in two decades.

There is much more to China and India than just being major "takers" who will continue to make major claims on available oil supplies. Although not openly stated, it may appear that both China and India had come to the same conclusion: that world oil supplies and access to these supplies likely may be constrained in the coming years and that efforts to secure needed supplies must begin now, not later. That justification is not universally shared, as internal planning documents of some companies may underscore an anticipated softening of the world oil market in the coming years as supply again overtakes demand.

MEASURING THE SUPPLY DIVERSIFICATION EFFORT

China's search for equity oil worldwide has caught the attention of the media, host governments, other national (state-owned) oil companies (NOCs) and, not surprisingly, international oil companies (IOCs). Some observers are inclined to view this search as an indicator of anticipated physical limitations on oil supplies. Others see it as competing with the United States and India for access. Neither point of view is correct. There is no overt evidence to support the contention that China is driven by a concern over future oil shortages. Competition for access to new sources of oil is an everyday occurrence, and care should be taken not to characterize China's search for equity oil as placing that country in direct competition with the United States.

In sum, China is conducting itself no different than any other oil-importing country as it seeks security and diversity of supply.[9]

The challenge to IOCs, also seeking access, is found in China's very aggressive efforts to pursuing overseas investment, with Chinese national companies overspending or promising too much just to secure a contract.[10] Economic concerns and commercial interests may not always be the drivers in their search for equity oil.

Although China is possibly prepared to pay a premium as a form of insurance against supply disruptions, competitors would argue that, for example, long-term contracts providing for the supply of natural gas delivered by pipeline, or as LNG, would offer the same protection. Others warn that the desire to own the assets in the ground could well raise the danger of disruption by depending on oil or gas from developing countries characterized by unstable or capricious regimes.[11] In today's world of oil, ownership of oil assets matters little. In other words, for China at least, diversity of supply is still the best protection.

State-owned Chinese oil companies seeking deals with state-owned companies of oil-exporting countries are changing the nature of the "oil game." One IOC in particular has risen to the challenge posed by China's national oil companies. Royal Dutch/Shell for one has advised oil-producing countries to be wary of signing deals with Chinese companies, saying to do so could expose them to interference from these governments.[12] Shell added that "government-to-government deals introduce a dependence (on the consuming nation government) and that government has its own agenda."

Striking deals with Central Asian countries—Kazakhstan and Uzbekistan—are examples where oil and political interests are both served. China wants access to their oil and natural gas, but China also wants regional stability, and that includes avoiding actions that might encourage separatism in the border region of Xinjiang, now coming onto its own as an oil producer.[13]

Nonetheless, the key question is this: how successful has China been to date, not in buying oil on the market for current needs, but in developing equity oil—that is, oil *owned* by China?

China's total investment in oil and natural gas projects outside the country has been estimated at $7 billion, covering more than 30 countries and regions around the world, involving 65 cooperation projects.[14] To date, this effort has found only mediocre success. In 2003, Chinese state-owned companies produced 220,000 b/d of equity oil. But the real successes await. The volume of equity oil is to rise by 8 percent (annually) through 2020 when equity oil hits 1.4 million b/d.[15] Nonetheless, the prices paid and agreed terms emerging from this "shopping spree" would fail to meet financial and commercial criteria acceptable to most IOCs.

The list of exporting countries where China currently has—or is seeking—equity oil is long, and continually expanding. The more relevant efforts are described in the following, not in any particular order of priority.

Angola

China rather fiercely protects its interests in Angola, characterized in large part by its willingness to pay what it takes to win, especially in its competition with India. China outbid India and acquired from Shell its 50 percent stake in BP-operated Block 18. China outmaneuvered India by offering a $2 billion oil-backed credit line for a variety of projects in Angola.[16] This oil-backed credit line—that is, the $2 billion loan will be repaid with oil—will be released on a project-by-project basis to rebuild Angola's war-damaged infrastructure. A mutual dependency is developing between China and Angola, and China already imports around 30 percent of Angola's oil.

Oil Pipeline from Angarsk (East Siberia) to Daqing

This pipeline, proposed by Mikhail Khodorkovsky, then president of Yukos, at that time Russia's leading oil company, fell victim to Khodorkovsky's jailing and the re-nationalization of Yuganskneftegaz, removing from Yukos its key performer. Moscow had not supported the concept of a privately owned oil export pipeline, with a carrying capacity of 560,000 b/d, and argued against the line serving just a single market. Is it a coincidence that Russia's rail system has suggested that the volume of oil it could move to China would equal the proposed pipeline carrying capacity?

The option to build a pipeline to Daqing has remained open, but is projected now as a branch off a pipeline to be built from East Siberia to the Pacific Ocean port of Nakhodka.[17] China still believes that Russia will give priority to this proposed branch line, in part because of

China's production payment of $6 billion to secure the delivery of 48.4 million metric tons of oil over a period of 5 years.[18] There is a hidden risk in all this that could work against Japan's national interests. Suppose the pipeline to Nakhodka is not built or construction is delayed. China would get the oil it needs, but Japan would not. It is unclear whether Russian rail would then step in to deliver oil by rail beyond China to Nakhodka, presuming sufficient exportable volumes were available.

In late May 2005 the Russian ambassador to China, in his departing address, said that Russia will prioritize China over Japan as the recipient of oil supplies from an oil pipeline that would link East Siberia with the Russian Far East.[19] In April 2005 Moscow had stated that an oil pipeline would be built from Taishet to the halfway point at Skovorodino, near the Russian-China border. If a geological survey finds crude oil reserves in East Siberia to be insufficient, the pipeline would not be extended to the port of Nakhodka on the Pacific Ocean. Conversely, if oil reserves warrant, the line would be completed as originally envisaged.

The Russian decision on the surface enhances Russian-Chinese relations, but damages relations with Japan. Does it matter to the United States whether a pipeline to Nakhodka, and a branch line to China, is built? If it does not, then it should.[20] The United States should first support a pipeline to the Barents Sea port of Murmansk, as it has, or to nearby Indiga, whichever is best suited for facilitating exports to U.S. markets. Conversely, a pipeline serving Nakhodka, with a branch China, and selling to Asia-Pacific buyers in general, likely would replace oil from the Persian Gulf. This displaced oil in turn would become available to U.S. markets, raising dependence on that region—perhaps not in the U.S. national interest.

Acquiring Unocal Corporation

Unocal had attractive assets in Asia that would meet Chinese interests and needs.[21] As early as February 2005, it was rumored that CNOOC was interested in acquiring Unocal but that the executive management decided against making a bid. CNOOC's interest may have given impetus to other companies who had been eyeing Unocal for some time, again for its Asian assets. At the same time, those prospective bidders also interested in the Chinese market were faced with the need for balancing these perhaps conflicting interests, not wanting to damage current or prospective business dealings with China.

CNOOC's late bid in June, however, triggered opposition from Capitol Hill, tied in many ways to other issues—trade practices, currency valuation, or an unfair playing field in regard to government funding of the bid. This "political environment" was cited as one of the main reasons that CNOOC withdrew its bid.[22] Ultimately, Chevron, which made an initial bid for Unocal in April 2005, was successful.

CNPC Interest in Yuganskneftegaz

Russian authorities forced the sale of Yuganskneftegaz, which had been the key producing company controlled by Yukos, ostensibly to cover unpaid back taxes. For a number of months it had not been fully clear what would happen to Yuganskneftegaz, as both Rosneft and Gazprom struggled for ownership. Rosneft emerged the winner and likely will take ownership of 1 million b/d producer Yuganskneftegaz sometime this summer.

Yugansneftegaz had been acquired at auction by a completely unknown buyer, Baikalfinancegroup, for just $9.35 billion, roughly one-half the company's value.[23] Rosneft then stepped in and purchased Baikalfinancegroup. Rosneft likely had to seek external financing to cover that purchase and may have secured a short-term loan to do so. Subsequent reporting revealed that Rosneft may have borrowed money from the government in what appeared as an insider deal.[24]

CNPC had been seen as a possible source of financing the purchase of Yuganskneftegaz by Rosneft, having extended to that company a $6 billion loan backed by future crude oil deliveries.[25] Russian authorities quickly denied the report, revealing that Rosneft had arranged a $6 billion prepayment from the China National Petroleum Corporation to cover the shipments of 48.4 million metric tons of oil to China between now and 2010.[26]

That agreed volume equates to the delivery of roughly 194,000 b/d, assuming that 48.4 million tons were delivered over a five-year period, 2006–2010. The oil in turn would come from other subsidiaries of Rosneft, not from Yuganskneftegaz.

Delivering roughly 194,000 b/d clearly is within the present capability of the Russian railway system. What is not clear, however, is whether that 194,000 b/d to be delivered to China is incremental to rail movements already agreed upon. More importantly perhaps, when considering the $6 billion prepayment, China paid about $17 per barrel, well below what the market normally would call for.

Russian energy minister Viktor Khristenko earlier had stated that CNPC might be invited to acquire a 20 percent stake in state-owned Yuganskneftegaz.[27] This comment was given on the same day that Prime Minister Mikhail Fradkov signed off on constructing a pipeline that would move oil from East Siberia to Nakhodka on the Pacific Coast. Nothing in his directive was said about building a branch oil pipeline to China. CNPC has rejected Khristenko's proposal, possibly holding out for a higher percentage stake.

Was the timing of these two statements coincidental, or was offering to China the prospect of an equity position in Yuganskneftegaz designed to repair any political damage resulting from then apparent loss of a pipeline link between Russia and China?[28]

Oil Pipeline from Kazakhstan

China currently imports very small volumes of oil from Kazakhstan. In 2003, for example, these imports averaged less than 23,000 b/d, up from about 19,000 b/d in 2002.[29] Delivery is by rail.

Volumes and means of delivery will change in the coming years. Construction of an oil pipeline from Atasu, in the Karaganda region of Kazakhstan, to Alashankou, in western China, was initiated in late September 2004. Upon completion this pipeline will be China's first major diversification among import routes. The pipeline is a joint 50-50 venture between a subsidiary of CNPC and Kazakh state-owned KazTransOil. About 1,000 kilometers in length, the pipeline is to be completed by December 2005, so the Kazakh government has emphasized, although that completion date may be unrealistic. The initial carrying capacity of the line will be 200,000 b/d, doubling to 400,000 b/d by 2011 with completion of the pipeline along its entire route—that is, following construction of all the planned pumping stations.[30]

The full cost of the 32-inch line will be about $3 billion. It will reach westward to link up with the existing Atyrau-Kenkiyak pipeline.[31] The fill for the pipeline will come in part from an investment that China has made in western Kazakhstan, where it has secured a 60 percent stake in Aktobemunaigaz. Additionally, some volumes of West Siberian crude oil could be placed in this pipeline.

Iran

Iran has expressed a desire to sell more oil to China, which currently looks to Iran for about 14 percent of the oil it imports, placing Iran second only to Saudi Arabia in importance as a foreign supplier. A number of oil- and gas-related deals, large and small, have been concluded with Iran. For example, CNPC purchased the Iranian subsidiary of Sheer Energy, a Canadian firm, giving it a 49 percent stake in the Masjed-I-Sueiman oil field, a seven-year arrangement worth $121 million.[32]

At the other end of the financial scale stands the $70 billion agreement under which Sinopec will purchase 250 million tons of LNG over a period of 25 years. In exchange, Iran will export 150,000 b/d of crude oil to China following development of the Yadavaran oil field, in which Sinopec will have a 50 percent stake.[33] Yadavaran will be developed under so-called buyback conditions that are generally rejected by IOCs because they do not offer sufficient return on funds invested.

With this growing oil trade, any effort by the United States to seek, through the UN Security Council, expanded sanctions on Iran—sanctions that possibly might interfere with oil flows between Iran and China—could be taken by China as a step by the United States to contain that country's growing political and economic strength and influence.

China, a member of the UN Security Council and valuing the Iranian oil it has come to depend upon, might not be willing to support any U.S.-backed initiative that would impinge on oil exports from Iran. The ultimate decision, of course, would reflect those actions that best served China's national interests at the time.

Venezuela

Venezuela is a key member of OPEC, both in oil production and oil exports. Venezuela currently claims production of 3.1 million b/d, but most observers accept 2.6 million b/d as a more realistic level. The larger share of production is exported, of which 1.521 million b/d were imported by the United States during 2004. Saudi Arabia, with sales to the United States of 1.556 million b/d, barely edged out Venezuela for third place among the U.S. sources of oil imports.[34]

Political tensions between the United States and Venezuela continue to rise. President Chavez has been touring the world, ostensibly in an effort to unite in some fashion all those states opposed to the policies of the United States. More importantly, Chavez has been buying military hardware in amounts clearly in excess of domestic requirements, including recent purchases from Russia.

In response the United States is developing a policy to "contain" the efforts of President Chavez to employ his oil income and influence to subvert countries where the social fabric is the weakest.[35]

A recently completed deal calls for 100,000 b/d of Venezuelan crude oil to be exported to China, together with 60,000 b/d of fuel oil, and also 36,000 b/d of Orimulsion boiler fuel.[36] The crude is a heavy crude, normally destined for the U.S. market.

Will arms sales authored by Russia and the commercial arrangements struck with China intrude on relations between Venezuela and the United States? The United States can be expected to object to the arms sales but not to any strictly commercial oil-related arrangements with other countries.

Oil Sands of Canada

China has confirmed its interest in the oil sands of Canada, both in participating in project development and perhaps in assuming an equity position in a pipeline that would carry the oil to a port of export on the British Columbia coast. This pipeline, proposed by Enbridge, Inc., would cost $2.5 billion and would have a carrying capacity of 400,000 b/d, with the majority of the oil carried destined for China.[37] Present Canadian oil sands output is about 1 million b/d, with plans to expand to 2 million b/d by 2010 and further to 3.3 million b/d by 2015.

Two oil sands newcomers, UTS Energy and Synenco Energy, have been looking for project partners and have indicated that Asian interests taking controlling stakes would not be ruled out. Sinopec confirmed it was exploring investment in Canadian oil sands.[38]

Sinopec was successful in its search and for C$105 million will acquire a 40 percent interest in Synenco Energy's Northern Lights project near Fort McMurray.[39] The project is valued at C$4.5 billion.

Start-up is envisaged for late 2009 or 2010, either in two phases of 50,000 b/d each or to come on stream at 100,000 b/d.[40]

China's interests in the oil sands of Canada had been confirmed in mid-April 2005 with the purchase for C$150 million of a 17 percent stake in MEG Energy Corporation, a closely held Calgary-based company. The oil sands project China bought into is comparatively modest in the short term, hoping to produce 25,000 b/d by 2008. Longer term, the project could expand to 145,000 b/d.

That purchase was followed shortly by an announcement that Enbridge had signed a deal to aggregate 200,000 b/d to make PetroChina the anchor tenant for its planned Gateway pipeline.[41] This pipeline would be laid from northern Alberta to a deepwater port either at Kitimat, British Columbia, or Prince Rupert for export to Asia and California markets. The pipeline will cover a distance of 720 miles and will have a carrying capacity of 400,000 b/d. The cost of construction has been placed at $C2.5 billion, with completion contemplated by 2010.

Canada, very much aware of its position as the leading foreign oil supplier to U.S. markets, would not want any oil sands–related contracts with China to disturb that mutual dependency or to intrude on relations in any way. The United States would, it is hoped, view any effort that led to development of additional oil supplies as mutually beneficial and as a benefit to the world oil market as a whole, not just to Canada and China. In that regard, it is quite likely that some portion of the Chinese equity oil would find its way to markets on the U.S. West Coast.

Sudan

China has had a comparatively long presence in Sudan and takes considerable pride in the contributions it has made to that country's development.

"We started with Sudan from scratch. When we started there, they were an oil importer, and now they are an oil exporter. We've built refineries, pipelines, and production." Thus spoke the deputy director of the West Asian and African Affairs division of China's Trade Ministry,[42] a trade official who perhaps inadvertently provided a glimpse into how China will be conducting itself in its worldwide search for oil. When asked about Sudan's human rights record, he did not respond directly, but only replied that "we import from every source we can get oil from." China's deputy foreign minister confirmed this approach when, in a recent interview, he reportedly said "business is business,"[43] clearly implying that Chinese companies are not likely to be encumbered by human rights or environmental concerns.

International oil companies, those of U.S. origin in particular, do not enjoy that freedom of action, as their actions are closely observed by nongovernmental organizations (NGOs) everywhere. Any failure to protect basic human rights or the environment is treated to a loud and long public airing. As China becomes more and more involved in the world oil sector, will they be given a free ride, ignored by the NGOs, or will the reverse occur—that is, will China ignore the NGOs?

China has made a major commitment in Sudan. CNPC is the major stakeholder in the Greater Nile Petroleum Operating Company that was producing around 270,000 b/d in early 2004, with output expected to reach 350,000 b/d by the end of 2004.[44]

If an adequate exploration effort were made, production could expand to much higher levels. If not, then production could be expected to decline very sharply after 2007. The government remains optimistic and expects production to reach 500,000 b/d in 2005.

Sudan provides China's oil exploration and oil services a base in Africa, and Sudan's oil fields are China's only major international discovery and production success to date.[45] China is anxious to guarantee the security of its investment in Sudan, given the ongoing civil war, and has deployed several thousand troops in the country to protect the oil export pipeline it helped build.

Sudan in turn looks to China for arms to fight the southern rebels. Chinese-origin tanks, military aircraft, helicopters, and small arms have found their way to Sudan. Just as China is the leading importer of oil from Sudan, so has China become the largest supplier of arms to Sudan.[46]

China, as noted, is the largest importer of Sudanese oil and hopes to keep it this way.[47] Some 6 percent of the oil that China imports originates with Sudan, with another 14 percent provided by Iran. Sudan and Iran need China just as much China needs Sudan and Iran. Politics and trade, votes in the UN Security Council, and the Chinese desire to diversify its imports of oil all come together, with a barrel of oil serving as the symbol of this growing interdependency.

Uzbekistan

In late May 2005, the president of Uzbekistan, Islam Karimov, announced the signing of a $600 million oil deal with China that envisaged a joint venture between Uzbekneftegaz and CNPC.[48] Investments would focus on 23 oil fields in the region around Bukhara and Khiva.

It is not the scale of this financial commitment that is of particular interest. Rather, it has been noted that China's willingness to conclude a high-profile deal at a time when Uzbekistan faces international isolation over human rights abuses is testament to Beijing's priorities in relations with Central Asia.[49] The Chinese government has little interest in human rights issues; access to oil, and economic growth, overrides all.

Russia

Because there are no oil pipelines linking China with Russia, at present all of the imported crude oil and petroleum products arrive by rail. China imported 159,000 b/d of oil from Russia during 2004, a 57.5 percent gain over 2003 imports of 101,000 b/d.[50] Small volumes arrive from Kazakhstan, also by rail.

The head of the Russian railways system underscored that oil shipments by rail were scheduled to increase to 200,000 b/d in 2005, and further to 300,000 b/d in 2007.[51] Beyond that, by 2010, the Russian railways reportedly will have the capacity to move up to 530,000 b/d.[52] This claim in large part is to demonstrate that delivery of increasingly large volumes of oil can be accomplished just as easily by rail as by pipeline.

Indeed, Transport Minister Igor Levitin has been quoted as saying that if the annual delivery of oil to the Asia-Pacific region (read China) remains under 600,000 b/d, then rail transport will be far more reasonable, but for volumes above 600,000 b/d, the use of pipelines would be more profitable.[53]

Prior to the end of 2004, Yukos had been the only Russian supplier of oil to China. But with the effective renationalization of Yuganskneftegaz, the core oil producer of Yukos, would rail shipments to China continue? Other players quickly seized upon the opportunity to replace Yukos and sell to the expanding Chinese oil market. Moreover, delivery schedules have been enlarged over time, again perhaps more politically driven, in an effort to be responsive to Chinese needs, but by rail and not by the desired pipeline.

Lukoil began rail deliveries to China in November 2004. State-owned Rosneft joined the list in February 2005, having stepped in to replace Yukos.[54] All want to be viewed as helpful in satisfying China's appetite for oil, hoping that this early relationship will be beneficial over the coming years as other oil-related opportunities arise.

TNK-BP, a joint venture between TNK, a Russian oil company, and BP, has told China that it would ship only crude oil, not petroleum products, to China, as it is not competitive to do so.[55] In fact, in their judgment, crude oil deliveries would be able to compete with oil coming out of the Persian Gulf only if rail shipping charges would be reduced.

The Russian rail system has done just that. To further bolster the position of rail, Russian Railways (known as RZD) has indicated it will charge $72 per ton ($9.86 per barrel) to ship the first 100,000 b/d of oil from Angarsk in East Siberia to the Chinese border. Then, the price will be cut to $47 per ton ($6.44 per barrel) on shipments up to 200,000 b/d, and further to $36 ($4.93 per barrel) for shipments between 200,000 and 300,000 b/d.[56]

Notes

[1] Wang Ying, "Spotlight Shone on Energy Conservation," *China View,* http://www.chinaview.cn, January 20, 2005. Energy consumption is to decline from 2.68 tons of coal equivalents per 10,000 yuan of GDP in 2002 to 1.54 tons of coal equivalents by 2020.

[2] David Stanway, "Eliminating Waste Is Crucial to China's Energy Industry," *Interfax-China,* May 13, 2005.

[3] Geoffrey York, "Demand China's Unquenchable Thirst," *Globe and Mail,* http://theglobeandmail.com, May 21, 2005.

[4] Ken Silverstein, "China Takes Lessons from California," *Issue Alert,* Utili-Point International Inc., July 7, 2004.

[5] Keith Bradsher, "Alert to Gains by China, India Is Making Energy Deals," *New York Times,* January 17, 2005.

[6] "India May Merge Oil Explorers, Take on Chinese Rivals," *Bloomberg,* January 17, 2005.

[7] Jehangir Pocha, "The Axis of Oil," *In These Times,* http://www.inthesetimes.com/site/main/article/the_axis_of_oil, February 3, 2005. The writer correctly noted

that what worries Western powers the most are the growing ties between China, India, and Iran.

[8] Huma Siddiqui, "Chinese Premier's Visit May Pave Way for Energy Tie-up," *Financial Express*, New Delhi, February 6, 2005.

[9] A professor at China's National Defense University has noted that China's "energy diplomacy" serves three purposes: meeting energy needs, sharing the risk with other oil-producing countries, and upgrading its stature in the international community. See Radio Taiwan International, "China Uses 'Energy Diplomacy' to Suppress Taiwan," http://www.rti.com.tw/English/, May 28, 2005.

[10] Kang Wu and Shair Ling Hun, "State-Company Goals Give China's Investment Push Unique Features," *Oil & Gas Journal* 103, no.15 (April 18, 2005): 20.

[11] Guy de Jonquieres, "Risky Business Going Global," *Financial Times,* March 7, 2005.

[12] "Shell Warns Oil Producers on Chinese, Indian Deals," *Fairfax New Zealand Limited,* http://www.stuff.co.nz, April 22, 2005.

[13] Asia News, "Beijing's Interests in Central Asia Grow," http://www.asianews.it, April 28, 2005.

[14] Sharon Wu, "China Invests $7 billion in Overseas Oil Exploration," *Xinhua Financial News,* May 26, 2005.

[15] Logan, "Energy Outlook for China," February 3, 2005.

[16] Jacinta Moran, "At the Wellhead," Platts *Oilgram News* 83, no. 6 (January 10, 2005): 3.

[17] Construction of an oil pipeline from Taishet, in Eastern Siberia, to Nakhodka was approved at the end of December 2004. This line, with a capacity of 1.6 million b/d, reportedly would cost $11 billion to build, an estimate likely to be exceeded by a large margin, if and when construction is initiated. Of the projected volume, 1.12 million b/d would come from oil fields in West Siberia and 480,000 b/d from fields in East Siberia. See Isabel Gorst, "Russian Plan May Include Crude Pipeline to China," Platts *Oilgram News* 83, no. 9 (January 13, 2005): 1. Transneft, the Russian oil pipeline monopoly, has stated that the line would be completed by 2008 ("Pipeline Ready by 2008," *Moscow Times*, January 13, 2005). The pipeline, in the eyes of Moscow, will facilitate exports to Japan and the United States. Will the required volumes be available? If West Siberian oil production cannot be expanded by the 1.12 million b/d—and that is questionable—then volumes now flowing west for export would have to be diverted to fill this new pipeline. Moreover, current oil production in East Siberia is minimal and falls well short of 480,000 b/d. All this infers the initiation of a rather massive exploration and development effort in East Siberia, an effort that would have to be immediately

successful. As a first step in attempting to match production to pipeline capacity, Moscow is preparing to auction off, for exploration and development, 38 oil and gas fields in Eastern Siberia. Most of these fields, however, are either relatively small or have been barely explored, meaning that success is not at all guaranteed. See Valeria Korchagina, "38 Siberian Oil and Gas Fields Slated for Auction," *Moscow Times,* February 8, 2005.

[18] "Russia to Give China Oil Supply Priority," *AP,* Beijing, March 14, 2005.

[19] "China to Get Oil Before Japan: Russian Envoy," *Japan Times,* http://www.japantimes.co.jp, May 20, 2005.

[20] U.S. Secretary of Energy Samuel Bodman, during a visit to Moscow in late May 2005, commented: "To the extent it is viewed as being best for Russia to build every pipeline to China, then that's what they could do," adding that "I respectfully suggest that there are better ways." See Erin E Arvedlund, "In Moscow, U.S. Strives for Greater Share of Oil Exports," *New York Times,* May 25, 2005.

[21] Dennis Berman and Russell Gold, "China National Offshore May Want All or a Portion of No. 9 US Oil Company," *Wall Street Journal,* January 7, 2005.

[22] See the CNOOC press release, "CNOOC Limited to Withdraw Unocal Bid," August 2, 2005.

[23] Catherine Belton, "The Money Trail Leading to Yugansk," *Moscow Times,* June 6, 2005.

[24] Ibid.

[25] Catherine Belton, "Rosneft Seeks $6Bln from China's CNPC," *Moscow Times,* January 19, 2005.

[26] Nadia Rodova, "CNPC Declines to Confirm $6-bil Payment to Rosneft," Platts *Oilgram News* 83, no. 23 (February 3, 2005).

[27] Peter Lavelle, "Yukos' Chinese Afterlife," United Press International, Moscow, December 30, 2004.

[28] There have been conflicting media reports regarding the prospect that a 15 percent stake in Yuganskneftegaz had been offered to India's Oil and Natural Gas Corporation (ONGC) by Rosneft. Rosneft has said there have been no concrete negotiations about the purchase or sale of a minority stake in Yuganskneftegaz. See Guy Faulconbridge, "Gazprom Takeover of Rosneft Delayed," *Moscow Times,* January 14, 2005, p. 5. In the interim, ONGC reportedly had been granted permission by the Indian government to negotiate for an equity position in Yuganskneftegaz for $2 billion. See Guy Faulconbridge, "India to Enter Yugansk Poker Game," *Moscow Times,* January 11, 2005. The bid, if extended, would come from ONGC Videsh, the international exploration arm of ONGC. Now, with Yuganskneftegaz to be owned by Rosneft, might CNPC and ONGC still be offered an opportunity to acquire limited ownership in that company?

[29] Ambo News, "Kazakhstan/China: Start Construction on Oil Pipeline," September 29, 2004. These exports have actually been provided by a Canadian firm, PetroKazakhstan. See Isabel Gorst, "CNPC Wants to Add to Kazakh Upstream Presence," Platts *Oilgram News* 82, no.193 (October 7, 2004): 1.

[30] Oleg Antonov, Oral Karpishev, "Kazakhstan Develops Infrastructure to Provide Oil Export Growth," *Itar-Tass*, Almaty, December 29, 2004.

[31] Xinhua News Agency, "Construction on Sino-Kazakh Oil Pipeline Starts," *Alexander's Gas & Oil Connections* 9, issue 21 (October 28, 2004).

[32] Borzou Daragahi, "China Goes beyond Oil in Forging Ties to Persian Gulf," *New York Times*, January 12, 2005.

[33] Kevin Morrison, "China and India Raise the Stakes," *Financial Times*, January 8, 2005.

[34] Canada, with deliveries of 2.118 million b/d, held first place, followed by Mexico which delivered 1.642 million b/d to U.S. markets.

[35] Andy Webb-Vidal, "Bush Orders Policy to 'Contain' Chavez," *Financial Times*, March 13, 2005.

[36] Steve Ixer, "Venezuela, China in Pact for Crude, Fuel Oil Supply," Platts *Oilgram News* 83, no. 24 (February 4, 2005): 6.

[37] Dave Ebner, "China's Oil Sands Role Tests US," *Globe and Mail*, December 30, 2004.

[38] Gary Park, "China Poised to Invest in Canada's Oil Sands," Platts *Oilgram News* 83, no.10 (January 14, 2005): 2.

[39] Tavia Grant, "China's Sinopec Buys Oil Stake," *Globe and Mail*, http://www.globeinvestor.com, May 31, 2005.

[40] Gary Park, "Third Chinese Major Takes Stake in Canadian Oil Sands Project," Platts *Oilgram News* 83, no.104 (June 1, 2005): 1.

[41] Gary Park, "PetroChina Agrees to Ship Synthetic Crude on Planned Enbridge Line," Platts *Oilgram News* 83, no.72 (April 15, 2005): 1. In other words, Enbridge will help China find sources of oil, in addition to its take from the oil sands project it has bought into, to make a total of 200,000 b/d moving through the line. The "deal" signed was just a memorandum of understanding, as reported elsewhere. There is often a great distance to be traveled between a memorandum of understanding and a signed agreement.

[42] Howard W. French, "A Resource-Hungry China Speeds Trade with Africa," *International Herald Tribune*, http://www.iht.com, August 9, 2004.

[43] Ibid.

[44] EIA, "Sudan Country Analysis Brief," http://www.eia.doe.gov/emeu/cabs/sudan.html, March 2005.

[45] See http://platfirm.blogs.com/passionofthepresent/2004/10/summary_of_chin.

[46] "Sudan's New Peace Deal," *Alexander's Gas & Oil Connections* 10, issue 5 (March 10, 2005).

[47] Gerald Butt, "Thirst for Crude Pulling China into Sudan," *Daily Star,* http://www.dailystar.com.lb/, August 17, 2004.

[48] "China, Uzbekistan to Ink $600 Mln Oil Deal—Media," *Reuters,* May 25, 2005.

[49] Andrew Yeh, "Uzbekistan Signs $600m Oil Deal with China, *Financial Times,* May 26, 2005.

[50]"China Reports 50 Percent Rise in Crude Oil Imports via Rail," *China Post,* http://www.chinapost.com.tw, February 21, 2005. Reuters, reporting out of Singapore, said rail deliveries totaled 7.95 mmt, or 159,000 b/d, for a gain of 57.5 percent. The text reflects the latter source.

[51] Peter Lavelle, "Analysis: Yukos' Chinese Afterlife," UPI, Moscow, December 30, 2004. Subsequent reports indicate that this 300,000 b/d schedule would be achieved in 2006.

[52] Isabel Gorst and Anna Shiryaevskaya, "Russia Says East Siberia to Produce at Least 1-mil b/d to Fill Line," Platts *Oilgram News* 83, no.16 (January 25, 2005): 2.

[53] ITAR TASS, "Russia-China Oil Pipeline Profitable if Traffic Exceeds 30 Mln Tons," March 21, 2005.

[54] *RosBusinessConsulting,* "Yukos to Deliver 250,000 Tons of Oil to China," www.rbcnews.com, February 25, 2005. In late May 2005, Lukoil indicated that it would export just 2 million tons (40,000 b/d) by rail to China during 2005, however, explaining that the high oil export duties had made it more profitable to refine the crude oil for the domestic market. See *RosBusinessConsulting (RBC),* "LUKoil Not to Deliver Oil to China," May 27, 2005.

[55] Angelina Lee, "TNK-BP to Ship Only Crude, Not Products, to China," Platts *Oilgram News* 83, no.43 (March 4, 2005).

[56] "RZD Cuts Fees to China," *Moscow Times,* May 25, 2005.

CHAPTER SIX

PERCEIVED THREATS, BUT NOT TO THE NORTH

"STRING OF PEARLS" STRATEGY

According to a report sponsored by the Office of Net Assessments and prepared for the U.S. Secretary of Defense, China is developing military bases and diplomatic ties to protect its oil shipments and strategic interests. The report noted that "China . . . is looking not only to build a blue-water navy to control the sea lanes, but also to develop undersea mines and missile capabilities to deter the potential disruption of its energy supplies . . . ."[1]

The report underscored China's belief that the U.S. military will disrupt China's energy imports in any conflict over Taiwan and that China sees the United States as an unpredictable country that violates others' sovereignty and wants to "encircle" China.

On March 14, 2005, the Chinese legislature considered the passage of a law that would authorize an attack on Taiwan should the island declare its independence.[2] Taiwan does not recognize Chinese sovereignty over the island, and China in turn has stated that the anti-secession law is necessary to protect its territorial integrity. The United States understandably opposes any unilateral effort to change the status quo. The law was passed by virtually unanimous vote.

The passage does not mean that military action against Taiwan is imminent.[3] It does confirm China's policy of seeking unification with Taiwan, independent from the mainland since 1949. The United States, though supporting a one-China policy and not Taiwan's independence, nonetheless spoke out against actions that raise tensions in the region.

Just how newsworthy was this vote? In October 2004 the National Intelligence Council released a publication, *Tracking the Dragon*, based on national intelligence estimates on China during the era of Mao, 1948–1976. It has been described as the largest single batch of declassified agency documents ever released. Special National Intelligence Estimate 13-3-61, in its conclusions, contained the following statement:

> Peiping has no compunctions about openly using its military forces to extend control when it can do so with little or no risk. It will continue its refusal to renounce the use of force for the seizure of Taiwan and the offshore islands . . . .[4]

With some concern the U.S. Department of Defense is watching the Chinese military buildup occurring partly with the purchase of equipment from Russia. It is not without reason that Secretary of Defense Donald Rumsfeld, in testimony before the House Armed Services Committee, observed that "[the Chinese are] increasingly moving their navy further distances from their shores in various types of exercises and activities."[5]

Why so? Crude oil imported from the Persian Gulf passes through the Strait of Malacca, one of the so-called chokepoints confronting worldwide oil tanker movements. Reportedly some 80 percent of the crude oil imported by China transits the Strait of Malacca.[6] The 600-mile-long strait between Indonesia, Malaysia, and Singapore offers the shortest sea route between the Gulf exporting countries and the importing countries of Asia. It is only 1.5 miles wide at its most narrow point, which creates the potential for collision, grounding, oil spill, piracy,[7] or terrorist attack.[8] Closure of the strait would immediately affect oil prices and would be followed by a time-consuming and expensive rearrangement of tanker flows.[9]

An estimated 11 million b/d of oil passed through the strait in 2003 en route to importers in Southeast Asia and the Far East. The importance of the strait will not diminish but rather will expand over time. It has been further estimated that some 20 million b/d will be transported by tanker through the strait in 2020, increasing to 24 million b/d by 2030 (see table 6.1).

China is very much aware of the vulnerabilities associated with a growing dependence on foreign oil and the reliance on sea routes. Indeed,

Table 6.1
Choke Points: Oil and LNG Movements through the Strait of Hormuz and Strait of Malacca, 2002 and 2030

	2002		2030	
Choke Point	**Commodity**	**Volume***	**Commodity**	**Volume***
Strait of Hormuz	Oil	15	Oil	43
	LNG	18	LNG	230
Strait of Malacca	Oil	11	Oil	24
	LNG	40	LNG	94

*Oil expressed in million barrels per day; liquefied natural gas (LNG) expressed in billion cubic meters.

Source: International Energy Agency (IEA), *World Energy Outlook 2004,* table 3.8, "Oil and LNG Tanker Traffic through Strategic Maritime Channels" (Paris: OECD/IEA, 2004), p. 119.

the Chinese Military Publishing House recently published a book—*Liberating Taiwan*—that imagines Chinese warships seizing sea routes to the Persian Gulf and imposing an oil embargo on Taipei, Tokyo, and Washington.[10]

How to reduce that vulnerability? China's shipping experts think that one approach would be for China to build its own oil tanker fleet, although the rationale behind this recommendation is not clear. Experts say that the fleet should be capable of handling at least 50 percent of China's total oil imports.[11] In their judgment, and based on the 50 percent share, the fleet should be capable of handling 1.5 million b/d by the year 2010 and more than 2.6 million b/d by 2020.[12] This proposal remains on paper; no concrete measures have yet been taken to build an oil tanker fleet. During the last several years, oil imports shipped by Chinese tankers made up only 10 percent of the total, while 90 percent were shipped by leased foreign tankers.

Japan, also heavily dependent on secure tanker movements through the strait, has urged Indonesia, Malaysia, and Singapore to do more to protect the oil tankers from piracy and terrorism. Now, another issue, equally serious if not more so, has emerged.[13] The tsunami that struck the region on December 26, 2004, may have brought about major changes in the physical configuration of the strait. For example, an unconfirmed report noted that one area of the waterway had its depth reduced to 105 feet from 4,060 feet. In another area, the depth was cut from 3,855 feet to just 92 feet. Making the strait safe for navigation will require not only considerable time but international cooperation as well. A more authoritative report, however, stated that the tsunami had not had any visible impact on the depths of the waterway.[14]

Another choke point—the Strait of Hormuz—provides transit for tankers carrying oil from Persian Gulf producers en route to consumers around the world. Should passage through that strait be denied, for whatever the reason, then an estimated 40 percent of world oil supplies would be affected. Could access be denied? The head of the Defense Intelligence Agency, in testimony before the Senate Intelligence Committee, warned: "We judge Iran can briefly close the Strait of Hormuz, relying on a layered strategy using predominantly naval, air and some ground forces."[15] No probability as to how long the oil flow might be disrupted was given.

It is the prospect of closure of the Strait of Hormuz or the Strait of Malacca that drives China to seek diversity among its suppliers of oil as well as diversity in how that oil would be delivered. Although China may believe that it can take steps to secure the Strait of Malacca, should the need arise, the Strait of Hormuz clearly lies beyond its control, beyond its influence.

WHEN CHINA LOOKS NORTH

When the United States looks north, it sees Canada, now the largest foreign supplier of crude oil and petroleum products to its southern neighbor. Equally important, Canadian gas exports provide one-sixth of the U.S. natural gas supply. Canadian sales of oil and gas to the United States serve the national interests of both countries. For the United States, Canada is a fully secure supplier, both share the same international border, and the United States will gladly import whatever volumes of oil and gas become available for export. For Canada, the United States represents what all oil-exporting countries seek—a nearby stable and growing market.

When China looks north it sees Russia, and the geologically promising but undeveloped oil and gas potential of East Siberia. Can a parallel be drawn between Canadian and U.S. energy relations and energy relations that might develop between Russia and China? Is it a given that oil and gas will flow south from Russia to China just as oil and gas have moved south from Canada to the United States? Is it in the national interests of Russia to serve as a supplier of oil and gas to China, and would closer political relations follow this resultant energy interdependence? Moreover, does the United States accept what may be inevitable and react accordingly?

Energy trade, as it develops between Russia and China, is and likely will continue to be a mix of politics and commercial relations. For China, as it is for all importing countries, it is a matter of access to sources of supply. For Russia, unless and until secure and substantial markets can be defined for the crude oil and natural gas resources known to be present in East Siberia, these resources will remain undeveloped, as they did for most of the second half of the twentieth century. Domestic requirements for oil and gas in East Siberia are much too limited, too scattered geographically, to warrant exploitation solely for that purpose.

Some observers see a deeper political rationale behind efforts to involve China and India in the Russian oil and gas sector: simply to give Russia leverage and support in the Yukos affair should a court challenge require it.[16] Although that leverage and support may not be needed, the promise of the China market as a catalyst for East Siberian development remains.

Notes

[1] "'String of pearls' Military Plan to Protect China's Oil: Report," *Khaleej Times*, January 18, 2005, http://www.khaleejtimes.com, referencing the *Washington Times* of Tuesday, January 18, 2005. The study, "Energy Futures in Asia," was actually prepared by defense contractor Booz Allen Hamilton and completed in late 2004.

[2] *Bloomberg*, "Rice Opposes Plan by China to Authorize Force against Taiwan," March 13, 2005.

[3] Bonnie Glaser, a senior associate in the CSIS International Security Program, when asked about the passage, had this to say: "Although this law does not signal a shift in Chinese policy towards Taiwan, it sets out several conditions under which force might be used against the island. The gap in perceptions of this law in Beijing and Taipei could not be greater. China views the law as reasonable and necessary; Taiwan claims it is a change in the status quo and a threat to the people of Taiwan. The passage of this law is unlikely to advance the cause of

peace and stability in the Taiwan Strait. Instead, it may well instigate new tensions." See *CSIS News Release*, "China's Anti-Secession Law," March 14, 2005, http://www.csis.org.

[4] *Special National Intelligence Estimate Number 13-3-61*, "Chinese Communist Capabilities and Intentions in the Far East," November 30, 1961. See National Intelligence Council, *Tracking the Dragon: National Intelligence Estimates on China during the Era of Mao, 1948–1976*, NIC 2004–2005, October 2004. The referenced SNIE can be found in the CD accompanying the publication. It is not included in the report itself.

[5] Eric Schmitt, "Rumsfeld Warns of Concern about Expansion of China's Navy," *New York Times*, February 17, 2005.

[6] Mengdi Gu, "China Wants More Pipelines for Improved Oil Import Security," *Oil and Gas Journal* 103, issue 1 (January 3, 2005): 59.

[7] The number of piracy attacks at sea fell in 2004 to 325, from the 445 attacks recorded in 2003, but the attacks were more murderous. See Frances Williams, "Fall in Global Sea Piracy Incidents," *Financial Times,* February 8, 2005.

[8] The International Maritime Bureau recorded 37 attacks in 2004 on vessels in the Strait of Malacca. See Jason Szep, "Security Lifted on Critical Waterway," *New Zealand Herald*, May 25, 2005.

[9] See Energy Information Administration (EIA), "World Oil Transit Chokepoints," April 2004, http://www.eia.doe.gov/emeu/cabs/choke.html.

[10] Brian Bremmer and Dexter Roberts, "Asia's Great Oil Hunt," *Business Week*, November 15, 2004. See also Wenran Jiang, "China's Quest for Energy Security," *The Jamestown Foundation*, December 2, 2004, p.1–3.

[11] "China Needs to Build Its Own Oil Tanker Fleet to Ensure Oil Security," *Alexander's Gas & Oil Connections,* May 10, 2005, http://www.gasandoil.com/.

[12] That implies Chinese oil imports will average 3 million b/d by 2010 and 5.2 million b/d by 2010. It is probable that these import levels are those developed by the International Energy Agency.

[13] Eric Watkins, "Japan Urges Malacca Strait shipping safety," *Oil and Gas Journal* 103, issue 4 (January 24, 2005): 28.

[14] Agence France Presse, "Malacca Strait Depth Altered after Tsunami," January 31, 2005.

[15] David S. Cloud, "U.S. Cites Iran Threat in Key Strait," *Wall Street Journal,* February 17, 2005.

[16] Chris Buckley, "Politics and Oil Mix in Russia-China Talks," *International Herald Tribune*, January 12, 2005.

CHAPTER SEVEN

CHINA'S NON-OIL ENERGY BASE

COAL: CHINA'S CONTINUING LIFEBLOOD

Current and future oil import levels and oil prices may attract the headlines of the day, but coal is still king in China (see table 7.1). Coal is scheduled to hold that position at least through 2020, then giving way a portion of its leadership first to natural gas and second to nuclear. Renewables will acquire a small relative gain.

Shifting the energy consumption balance away from coal and emphasizing other fuels will be a long-term assignment, as table 7.1 clearly illustrates.

Consumption patterns had changed little in 2004 from those of 2000, underscoring the difficult task ahead. Indeed, the share of coal had actually increased, as production was expanded to meet the needs of electric power stations as they struggled, unsuccessfully, to keep pace with demand.

China leads the world in terms of coal production and consumption of coal. Coal fuels almost 75 percent of China's electric power plants, with hydroelectric facilities providing 20 percent and oil and nuclear supplying the remainder.

Interestingly, while the United States is continually criticized for the fact that its demand for oil accounts for almost one-quarter of world oil consumption, the burning of coal in China by industry, households, and other uses makes up a reported 27 percent of worldwide consumption. At the same time, China is open to criticism for the environmental damage caused by the burning of low-quality coals and for the unacceptably high loss of life caused by mining accidents.

Table 7.1
Net Energy Consumption in China, 2002 and 2020
(percent of total)

	2002	2020
Coal	64.6	61.0
Oil	24.5	25.9
Natural gas	3.0	4.7
Nuclear	0.5	1.7
Renewables	7.2	6.4

Source: U.S. Department of Energy, Energy Information Administration (EIA), *International Energy Outlook 2005*, p. 147.

China may lead the world in terms of tons of coal produced, but China also leads the world in deaths caused by mining accidents.[1] There can be no more damaging indictment of the Chinese coal sector than the loss of life attributable in large part to ignoring acceptable safety practices. The sector itself is poorly regulated. Many of the accidents occur at small, unregulated and/or illegal mines where safety is generally ignored but where coal mining is often the only opportunity for employment.

More than 6,000 miners perished in mining accidents in 2004, or 80 percent of all global mining fatalities that year, but down from the roughly 6,400 lives lost in 2003.[2] Mine accidents in the first three months of 2005 killed 1,113 workers. A coal-mine blast on February 14, 2005, that at last report took 209 lives was the worst in decades.[3] Whether this accident leads to stricter observance of safety measures remains to be seen.

Nonetheless, China will stress that the number of deaths per million tons of mined coal had fallen from 5.07 in 2001 to 3.1 in 2004. The death rate in Chinese mines still is 100 times that of U.S. mines.

This loss of life is not offset by high productivity. Indeed, the average tonnage per Chinese miner was just 2.2 percent of the U.S. average, where many of the mining operations are mechanized.

As demand for coal continues to rise, reflecting economic growth in the country, many of the small, unregulated mines, once shut down, have now been returned to production. Just how much coal these mines provide is uncertain, which can be taken to mean that national coal production and consumption statistics are subject to question.

China recently passed a new law that closes, on safety grounds, those mines producing 90,000 tons or less a year.[4] It is questionable whether this law can be enforced. Pressure to mine more coal to meet the expanding requirements of industry, and the economy as a whole, for electricity is difficult to ignore. Typically, illegal mines are shut down following a major accident, but then the inspectors leave, and the mines reopen.

Much of the coal burned by industry and households alike is unwashed, high-sulfur coal that discharges tremendous amounts of sulfur dioxide into the atmosphere. China lags dramatically in installing clean coal-burning technology. The Asian Development Bank found that in 2003 just 5 percent of China's power plants had any serious pollution controls at all.[5]

Another approach to cut coal consumption, and in turn the mining of coal, would be to shut down power plants that are being built or planned without the needed approval of central authorities.[6] It has been estimated that China's unauthorized electric power capacity approaches 80 to 140 gigawatts. By shutting down these illegal plants, pressure on coal supplies hopefully would be eased. As always there is a tradeoff, and the tradeoff would be continued power shortages.

The mining of coal in China during 2005 is to provide the market with a bit more than 2 billion tons, presumed adequate to meet basic needs but yet not sufficient to offset inefficient production, overburdened transportation means, and an irrational distribution system.[7] As a consequence, the government anticipates continued coal shortages during the year. Coal supply had reached 1.9 billion tons in 2004 (table 7.2), but not enough to prevent regional shortages, with transport bottlenecks slowing up delivery between mines and consumers. A senior Chinese official was quoted as stating that "the present size and scale of China's coal industry are far from being able to meet the country's future market demand. Insufficient supply will continue to be a major problem."[8]

Coal's pricing system is divided into two parts.[9] To ensure electricity generation use, a certain amount of coal is ordered nationally. To meet the price of electricity, which has been set at an artificially low price, the price of these coals must also be kept low. The price for coal bought and sold on the market rises or falls in accordance with market forces.

Electric power generation will continue as the dominant consumer of Chinese coal, having contracted for delivery of 422 million tons of thermal coal in 2005.[10] Total consumption of coal during 2005 in electricity generation is expected to reach 1.15 billion tons. That implies, in turn, that the volume of coal bought on the market will roughly total 700 million tons.

As noted in table 7.2, only a minor increase in coal mining has been forecast for 2005, but the China National Coal Association has projected that demand would reach 2.1 billion tons, yielding a likely supply gap of 100 million tons. China's demand for electricity during 2005 is expected to rise by about 13 percent to 2.5 trillion kilowatt hours (kWh).[11] Shortages will continue, with these shortfalls reaching as much as 23,000 megawatts (MW).[12] Unfortunately, continued high rates of growth in power consumption beyond those initially forecast have resulted in much higher anticipated shortfalls.

It is a shortage of coal, not of oil, that confronts China as the year 2005 unfolds. The coal supply shortage reflects continued rapid growth in demand for electricity, with this demand outstripping the ability of the coal sector to respond. Coal mines are operating at near or even beyond design capacities. Despite continuing increases in coal production, supporting the generation of electricity, power shortages persist, countrywide. Although some relief probably could be obtained by raising prices, thus providing the consumer with an incentive to conserve, that step has not been taken. Electricity prices will not be raised nationally.[13]

Shortages could predominate for the remainder of this decade, with an anticipated shortfall in supply by 2010 equal to roughly 15 percent of production that year, but that would imply complete failure by the government to contain both the supply and demand for coal and electricity consumption. Demand for coal by 2020 has been placed at 2.6 billion tons, reflecting a gain on the order of 30 percent compared with demand in 2005. Can the coal industry respond, and does this demand level imply little or no improvement in air quality and little gain in coal substitution?

Although coal shortages currently preoccupy Beijing, Chinese officials nonetheless worry that in the not-too-distant future coal shortages could well be replaced by coal surpluses. Investment in coal mines is overheating, as investors are betting that high coal prices will be around

Table 7.2
Coal Production in China, Selected Years, 1990–2005

Year	Billion Metric Tons
1990	1.08
1992	1.12
1994	1.27
1996	1.40
1998	1.30
2000	1.19
2002	1.38
2003	1.48
2004	1.90
2005 (est.)	2.00

Sources: For 1990–2003, see EIA, *International Energy Annual 2003,* table 2.5: World Coal Production 1980–Present (table posted June 13, 2005), http://www.eia.doe.gov. Data in the EIA table are expressed in short tons and have been converted to metric tons, using the ratio of 1 metric ton equals 1.1023 short tons. For 2004–2005, see "Coal Shortage Becomes China's Top Concern in 2005 Economic Control," *Xinhua People's Daily Online*, January 27, 2005, http://english1.people.com.cn/200501/27/eng20050127_172145.html.

for some time.[14] Many of the new mines have ignored the necessary government approval before initiating construction. Indeed, capacity expansion goes against the government's plan to gradually reduce the output of small, inefficient mines. The director of the China Coal Industry Development Research Center has noted that output at these mines should be capped at the current 750 million tons per year and cut to 600 million tons by 2010.

If, in the face of continuing high world coal prices, China was unable to manage coal mining in the desired way but could reduce its demand for electricity, it might well emerge as an important supplier to the world coal market.

In looking for ways to cut oil imports, Chinese officials see another application of its world-leading coal reserves: the development of a coal liquefaction sector. The country's first such project is expected to begin operating in 2007, with an initial liquid output of 20,000 b/d. It is estimated that by 2013, 10 percent of oil imports will have been replaced by coal-liquefied oil.[15] Scaling up to that level is likely overly ambitious.[16] Coal liquefaction projects are very sensitive to two prices—the price of coal needed for the liquefaction process and the world oil market price. The key will be found in the ability to reduce liquefaction costs.[17]

NUCLEAR POWER IN CHINA

Security through diversity does not begin and end with oil. Rather, it extends to diversity among the types of primary energy available, which means giving more attention to nuclear power (table 7.3). Nor is nuclear power expansion in Asia limited just to China. Of the 27 units under construction worldwide in 2004, 16 were to be found in India, Japan, South Korea, Taiwan, and China. Moreover, 22 of the most recent 31 reactors to be commissioned are in Asia.

The rationale for more nuclear power plants extends beyond energy diversity. First, nuclear plants would replace coal-fired electric power, leading to improved air quality. A second priority attached to the Chinese nuclear program is simply to relieve electricity shortages, particularly along the east coast of the country.

Electricity shortages were prevalent throughout China during most of 2004 and continue throughout 2005. Realistically, however, the time frame required for nuclear reactor construction is far too long for this approach to be regarded as useful in mitigating current and likely near-term power shortages.

The establishment of a nuclear power industry in China began in 1954; the first nuclear plant, at Daya Bay, went operational in 1991.[18] By July of 2004 China had 9 nuclear power plants in operation, with a total capacity of 7,010 MW, supplying 48.3 gigawatt hours (GWh) of power (table 7.3). The Tianwan plant, under construction, is scheduled for initial operations in 2005, raising installed capacity to 9,130 MW.[19]

Construction of the country's largest nuclear power plant, at Yangjiang, is to begin before 2006. The plant will include six generating units, each with a capacity of 1 million kilowatts (kW).

A site for the fourth nuclear power plant in Guangdong Province will soon be selected, with construction to begin before 2010. Two nuclear plants already operate in the province at Daya Bay and Ling'ao. The total installed capacity of both plants is 4 million kW.

If construction plans are realized on schedule, China expects to have an installed nuclear power generating capacity of more than 36 million kW by 2020, providing 4 percent of total electricity generating

Table 7.3
China's Nuclear Power Sector Capacity, Selected Years, 1997–2050

Year	Capacity (MW)
1997	2,200
2004	7,010
2005 plan	9,130
2020 Plan	36,000*
2050	150,000

Source: Elizabeth A. Martin-Alldred and Robert J. Hard, "China: Romancing the Dragon," Nukem Inc., September 1997, p. 6.

*Subsequent reporting has raised the 2020 goal to 40,000 MW. See Reuters News Service, "Westinghouse Upbeat on China Nuclear Contract," Planet Ark, http://www.planetark.com/, May 23, 2005.

capacity that year, up from 2.3 percent of the national total in 2004.[20] To reach that goal, China plans to build at least 28 new reactors, which of course has attracted the attention of vendors worldwide.

To put the plans for expansion of nuclear capacity in the proper perspective, China last year had 440,000 MW of electric power generating capacity available. By the year 2020 that capacity is to reach to about 900,000 MW.

For comparison, 20 percent of U.S. electricity is provided by 103 nuclear power reactors, a higher reliance than that of Russia, for example, where nuclear electric power represents 16 percent of total power generated. In perhaps a more meaningful comparison, India secured 3.3 percent of its electricity from nuclear plants in 2003.

Will China become a major player in the world nuclear power industry if it is successful in reaching the stated plan for the year 2020? Not by any realistic measure, but then that is not the goal of the government. The goal is to provide diversity, to the extent it can, to its primary fuel balance, to reduce air pollution, and to try to balance electricity supply and demand.

In the interim, however, China plans to build reactors on a scale and pace comparable with the most ambitious nuclear energy programs the world has ever seen.[21] Given the limited opportunities worldwide for the construction of nuclear power plants,[22] competition for those contracts to be handed out by China is particularly intense. Three companies are in the running for the right to design and build the first 4 of more than 20 (some have estimated that the total would reach 27) new nuclear reactors to be built: Westinghouse (United States), Areva (France), and AtomStroiExport (Russia).[23] All three companies are drawing heavily upon the support of their respective governments.

It can be expected that China will be "going to school" during the actual construction process so that international participation in the nuclear reactor program can eventually be dispensed with, replaced by Chinese engineers, labor, equipment, and supplies.

As part of its nuclear power program, China is drawing up plans to construct and operate a pebble bed nuclear reactor for the first time in a commercially operated electric power station. It hopes this reactor, which at 195 MW will be just one-fifth the size of a standard nuclear reactor, will begin generating electricity in 2010.[24] The technology for this planned reactor will come from what China claims is the world's only test pebble bed reactor, which is located in a military zone outside Beijing.

Quick to take exception to China's claims, South Africa noted that it had a "three-to-four-year advantage" over its closest competitors in the race to develop a commercially viable fourth-generation nuclear power plant.[25] Nonetheless, construction of a pilot nuclear pebble bed modular reactor in Western Cape has been held up because of environmental concerns.

Scaling up of a pebble bed reactor represents a formidable task, and the cost per kW is roughly comparable to today's reactors. Advocates note that the reactor, imminently safe in their judgment, will not require a containment vessel. Critics in turn point out that the lack of a containment vessel makes the reactor an attractive target for terrorist or military attack. Moreover, the design does not solve the problem of nuclear waste, an issue of continuing concern to the United States.

NATURAL GAS

Natural gas does not yet play a major role in China's energy supply and demand. For the years 1990 to the present at least, domestic supply, very limited in itself, defined those volumes available for consumption. In 1990, domestic production averaged just 14.4 bcm, rising to about 40.8 bcm in 2004, 14 years later (table 7.4). Future growth in supply is to be based in large part on imports by pipeline and in the form of LNG.

The IEA forecasts a rising but not necessarily worrisome dependency on foreign supplies of natural gas, both in absolute and relative terms. This dependency, rising from zero in 2002 to 15 percent in 2010, is judged in that year to involve the import of 9 bcm. A steady growth, to 27 percent reliance, reflecting the importation of 42 bcm, by 2030 is forecast. Nonetheless, this expansion occurs over a period of 20 years and thus should not particularly influence the world gas market.[26]

That said, an anticipated growth in demand for natural gas reflects both environmental concerns and again the need for fuels diversification. Given the limitations of the domestic natural gas resource base, the question arises as to just how a balance between supply and demand might be achieved.

On the basis of the IEA forecasts, gas supply in 2010 can be placed at 60 bcm, growing to about 156 bcm by 2030. Importantly, of the implied increment in supply, imports are to account for roughly one-third, meaning that domestic production is expected to supply more than 110 bcm in 2030. That is probably an overstatement of the future performance of the Chinese domestic natural gas sector.

A Sinopec official recently presented its view of the future role of natural gas in China, using low-case and high-case scenarios (table 7.5). These broad ranges reflect uncertainties not so much with regard to demand, but probably more so with regard to natural gas supply availability. It can be presumed that the Chinese economy will consume whatever volumes of natural gas become available. Will development of the Kovykta gas field in Eastern Siberia move ahead? When? What volumes might be supplied to China? Will all the LNG projects come online as hoped for, or will there be delays? Absent hoped-for natural gas supplies, planners and the economy as a whole most likely would look to the coal sector as a substitute, placing further strain on that sector. Pollution abatement derived from the greater use of natural gas would be lost.

West-East Natural Gas Pipeline

China has decided on a number of differing approaches to balancing supply and demand for natural gas. The first approach centers on the

Table 7.4
Production of Natural Gas in China, Selected Years, 1990–2004

Year	Billion Cubic Meters
1990	14.4
1995	17.0
2000	27.2
2002	32.6
2003	34.3[a]
2004	40.8[a]

Source: Except where otherwise noted, from Department of Energy, "An Energy Overview of the People's Republic of China," http://www.fe.doe.gov/international/EastAsia_and_Oceania/chinover.html; *BP Statistical Review of World Energy,* June 2004, p.24.

[a] Vandana Hari and Winnie Lee, "China's Oil Demand Jumped 15 Percent Last Year," Platts *Oilgram News* 83, no. 36 (February 23, 2005): 2.

Table 7.5
Future Natural Gas Demand in China, Selected Years, 2005–2020

Year	Range (billion cubic meters)
2005	49–64.5
2010	70–112.1
2015	90–150
2020	150–250

Source: Kim Wong, "China Seen Sharply Hiking Gas Use This Year," Platts *Oilgram News* 83, no. 38 (February 25, 2005): 4.

West-East natural gas pipeline, completed for operation at the end of 2004. This 4,000 kilometer pipeline will move natural gas from Lunnan in the Tarim Basin to Shanghai. Built at a cost of $8 billion, it handled a bit more than 1.3 bcm in 2004 and is scheduled to transport 4 bcm in 2005. Deliveries are to reach the pipeline designed capacity of 12 bcm by 2007.[27] These are comparatively small volumes, given the length of the pipeline, and probably represent the capability of the gas fields linked to the pipeline.

Gas Supplies from Kovykta

The second approach envisages construction of a pipeline linking the Kovykta gas field, located in East Siberia, to the north and west of

Irkutsk, with consumers in China and Korea.[28] Kovykta has been a subject of interest for some time now. At present, negotiations relate to the routing of the proposed line and to the role to be played by Gazprom. Gazprom controls all pipelines exporting Russian natural gas to buyers outside the country and is unlikely to stand aside while Kovykta is being developed or to be accepting of an export pipeline that is not under its control. Present thinking envisages the proposed pipeline bypassing both North Korea and Mongolia. The first deliveries of 20 bcm of Kovykta gas to China were to begin in 2010, with 10 bcm directed to South Korea. That timetable would appear to be somewhat optimistic, given the absence of any supporting infrastructure, the distance between supplier and consumer, the difficult terrain to be crossed, and the bureaucratic hurdles that lie ahead. There is as yet no formal agreement to move ahead.

In the interim, Moscow has been threatening to take away TNK-BP's rights to develop Kovykta for failure to move ahead with development plans.[29] TNK-BP has responded by approving the expenditure of $135 million for the design and construction of what it calls the "Early Gas" phase of the Kovykta Project.[30] These funds will be used to supply pipeline gas to customers in nearby areas. This regional gasification project is to supply 2.8 bcm by 2020.

This limited response may not be enough. Roughly one week following TNK-BP's announcement, the *Moscow Times* reported that pressure appeared to be mounting again on TNK-BP and that the Kovykta license was under review, with a final decision to be made in June.[31] To keep the license, what will TNK-BP have to do? Nothing more than in some fashion cede control to Gazprom, and it has done just that, proposing to Gazprom that the two organizations engage in a joint venture to develop this huge gas field.

Kazakhstan-China Natural Gas Pipeline

Another element of this second approach relates to the prospect of building a natural gas pipeline from Kazakhstan to China. Chinese authorities have indicated a need for 8 to 10 bcm of Kazakh natural gas by 2008 and 30 bcm in 2020.[32] Negotiations continue and if concluded successfully, delivery of Kazakh gas to China could begin in late 2008. The United States has indicated it would support this proposed pipeline.[33] For Kazakhstan, multiple pipelines in multiple directions, while minimizing the role of Russia as a transit country, serve its national interests.

Liquefied Natural Gas (LNG)

The third approach relates to the importation of LNG primarily to serve customers along the southeastern coastal region. A variety of suppliers have been investigated, and supply contracts have been signed with Australia, Iran, and Indonesia, with the gas in large part to be burned in the generation of electric power, substituting for oil. First deliveries of LNG, from Iran, are to begin in 2009.

Coalbed Methane

A fourth approach to covering future demand for gas recognizes the considerable potential—one not newly discovered—of coalbed methane. A number of years ago in-place coalbed methane resources in China were estimated at 30–35 trillion cubic meters (tcm),[34] although a more recent report[35] placed the coalbed methane potential at about 22.5 tcm. Nonetheless, coalbed methane is neither easily nor quickly developed, in part because of the large number of wells that must be drilled to exploit its potential. Current goals, if achieved, would provide about 2 bcm of coalbed methane by 2010, for only a very minor contribution to supply if realized.[36]

Natural Gas Hydrates

Finally, natural gas hydrates have attracted the attention of Chinese researchers. The scientific community for years has been aware of the almost immeasurable potential of gas hydrates, but competitive mining of these hydrates rests beyond known and acceptable technology. Hydrates are commonly formed from natural gas and seawater. There is only one gas hydrate field producing commercial gas volumes in the world. That field, Messoyakha, located in Russia's East Siberia, has been producing for 35 years, and its gas is transported by a 263-kilometer pipeline to the city of Norilsk, where it is burned to generate heat and electricity.[37] Production costs are 15 percent to 20 percent higher than other fields in the region because of special technology requirements.

RENEWABLES

Renewable forms of energy—hydroelectric power, wind, solar, geothermal, and biomass—do not in themselves contribute substantially to national energy supply, though local and even regional importance is unquestioned. Renewables supplied 7.2 percent of total energy consumption in 2002 (see table 7.1).[38]

China acknowledges the need to counter the nation's worsening pollution problems. As an example, about 60 percent of China's 768 million rural residents still cook over open fires, contributing to poor air quality.[39] How to eliminate chronic energy shortages and counter the increasing reliance on imported oil and natural gas? Are there other forms of domestically available energy that could be called upon?

At the end of February 2005 China passed a Law on Renewable Sources that calls for raising the consumption of renewables to 10 percent of total national energy consumption by the year 2020.[40] This new law, to come into force in 2006, offers a number of incentives to encourage the availability and consumption of renewables, including lending and tax incentives and the requirement that power grid operators buy in full volume that output from registered energy producers (read electricity) within their domains.[41]

Additionally, an energy conservation law and a regulation on building energy conservation have been on the books for at least four years.[42]

The initial reaction to this goal of 10 percent renewables by 2020 has been negative. For example, wind-generated electricity currently contributes just 0.01 percent to the power grid. A recent assessment attributes to China a potential for more than 100,000 MW of renewal power—that is, including both solar and wind.[43]

Hydropower is China's second largest energy resource next to coal. Some sources estimate China has the potential for about 400 million kW of hydropower, yet less than one-fourth of that potential has been utilized.[44] Much of the greatest potential is in the southwest, far from population centers.

China is seeking to exploit this potential in the coming years. The Three Gorges Dam project, when completed in 2009, will provide more than 18.2 GW. There are also 11 major hydro projects planned for the upper portion of the Yellow River with a total capacity of 15.8 GW. Another huge-scale project, in the planning stages for the Jinshajiang tributary of the Yangtze, will, when completed in 2009, have generating capacity nearly double that of Three Gorges.[45]

China understands that it needs access to the latest renewable energy technologies and renewable energy equipment, such as solar cells and wind turbines. Liu Jang, vice chairman of the National Development and Reform Commission, has urged that Western countries give Chinese industry access to these technologies and to manufacture renewable energy equipment in China.[46] High fees for technology transfer, he added, have held back the application of these technologies worldwide.[47] He reasoned that forced utilization of obsolete technologies would lead to severe climatic consequences.

The U.S. Department of Energy is responding in part to this request, having established a U.S.-China joint working group that has identified 10 areas for cooperation, including natural gas technology, combined cooling, heating and power, clean coal, hydrogen and fuel cell vehicles, and solar photovoltaics.[48] This effort is designed to help China keep its promise that World Health Organization standards for urban air quality will be met by 2008, in time for the Beijing Summer Olympics.

Nonetheless, recognizing that action needs to be taken and that ways must be found to minimize the myriad of energy problems confronting the country is perhaps as important as the indicated goal itself. Costs will be extremely high, possibly as much as $157 billion for future environmental protection. Moreover, China calculates that only when investment in environmental protection reaches 3 percent of GDP can a country improve its environment noticeably.

Notes

[1] Brian Rhoads, "China Mine Blast Kills over 200," *Reuters*, Beijing, February 15, 2005.

[2] Mure Dickie, "World's Riskiest Mines Claim 200 Lives," *Financial Times*, February 16, 2004.

[3] "Rescue Work Continues after Coal Mine Blast," *China Economic Net*, February 16, 2005.

[4] Richard McGregor, "China's Coal Entrepreneurs Tap a Rich Seam," *Financial Times*, January 17, 2005.

[5] Brian Bremner, Dexter Roberts, with Adam Aston, Stanley Reed, and Jason Bush, "Asia's Great Oil Hunt," *Business Week*, November 15, 2004, http://www.businessweek.com.

[6] "Illegal Power Plants to be Cracked Down," *China Economic Net*, February 16, 2005.

[7] United Press International, "China Faces Coal Crunch in 2005," Beijing, January 4, 2005.

[8] Elaine Kurtenbach, "Report: China Facing Coal Shortage," *Business Week*, http://www.businessweek.com, May 25, 2005.

[9] "Market-Orientation Is the Key for China's Coal Industry Reform, Experts," *People's Daily Online*, April 18, 2005, http://english.peopledaily.com.cn/200504/18/eng20050418_181604.html.

[10] "CNOOC Looks to LNG as Future," *China View*, http://www.chinaview.cn, January 28, 2005.

[11] "Power Demand to Rise 13 Percent," *Shenzhen Daily*, http://www.ce.cn, *China Economic Net*, March 3, 2005.

[12] Lindsay Beck with additional reporting by Judy Hua, "Booming China Struggles to Ease Power Shortage," Reuters, Planet Ark, http://www.planetark.com, February 28, 2005.

[13] "China's Electricity Shortage to Relieve This Year," *China Economic Net*, February 25, 2005. This headline is misleading. An official of the State Electricity Regulatory Commission stated that the shortage would remain unchanged.

[14] Xinhua News Agency, "Exploding Energy Sector Risking Glut," May 26, 2005, http://news.xinhuanet.com/english.

[15] "Coal Liquefaction to Ease Oil Import Burden," *China Economic Net*, January 24, 2005, http://www.en.ce.cn/subject/EnergyCrisis/ECchina.

[16] A senior official of Synfuels China, which is part of the Institute of Coal Chemistry, sees viably priced output reaching 200,000 b/d to 600,000 b/d within a decade. See "Solution to Oil Supply Shortage May Lie in China's Coal Reserves," *China Daily*, http://en.ce.cn/Industries/Energy&Mining, *China Economic Net*, May 26, 2005.

[17] China holds one-fourth of total world shale oil resources. These resources are placed at about 31.6 billion tons. China may partner with Shell Exploration to exploit these resources, using Shell's in situ conversion process. See "Jilin, Shell Team up for Oil Shale," *China View*, http://www.chinaview.cn, January 11, 2005.

[18] A National Intelligence Estimate (NIE 13-2-60), entitled "The Chinese Communist Atomic Energy Program," had concluded that although it had been announced in 1956 that "atomic power stations would be built," such stations

were not included in the Second Five Year Plan (1958–1962), and there was no present evidence for a power program. It was estimated that the Chinese would not construct nuclear power stations in the 1960–1965 period. See National Intelligence Council, *Tracking the Dragon*, p. 308.

[19] China Internet Information Center, http://www.china.org.cn/english/2004/Sep/105814.htm.

[20] Mure Dickie, "China Sees Role for Foreign Suppliers in Meeting Atomic Energy Targets," *Financial Times*, September 2, 2004.

[21] Howard W. French, "China Promotes Another Boom: Nuclear Power," *New York Times*, January 15, 2005.

[22] Only one reactor order has been placed outside Asia—by Finland in 2003—since the accident at Chernobyl. There have been no new orders in the United States since 1978.

[23] Jasmine Yap and Nicolas Johnson, "Areva, Westinghouse Chase China Nuclear Deals amid Global Slump," *Bloomberg*, March 2, 2005.

[24] Mure Dickie, "China in Drive For Revolutionary Reactors," *Financial Times*, February 8, 2005. See also Mure Dickie, "China Set to Pioneer Meltdown-Proof Reactor and Take Lead in Nuclear Race," *Financial Times*, February 8, 2005.

[25] Dave Mars, "SA Nuclear Company Claims 'Four-Year Edge' on Chinese Rivals," AllAfrica Global Media, http://www.allafrica.com, February 9, 2005.

[26] IEA, *World Energy Outlook 2004,* table 4.2, p 140. As noted, in absolute terms, imported gas is projected to increase from 9 bcm in 2010 to 42 bcm in 2030. Surprisingly, the absolute growth in imports of natural gas by India during this same time period—34 bcm—exceeds if only slightly the growth in imports by China.

[27] "China Gas Pipeline May Earn a Profit in 2005," *Alexander's Gas & Oil Connections*, January 27, 2005.

[28] Russian planners see East Siberia, with known reserves of 6.6 trillion cubic meters, as rapidly becoming the gas center of Russia, with regional production in 2020 reaching 110 billion cubic meters per year. See *RFE/RL,* "Russia, China, and the Politics of Energy," January 5, 2005.

[29] TNK-BP holds 62 percent interest in RUSIA Petroleum, the license holder for Kovykta. Interros holds 26 percent of RUSIA, and the Irkutsk region's administration holds 12 percent.

[30] Interfax Information Services, B.V., "TNK-BP allocates Funds to Develop Major Gas Field in Siberia," *Rigzone*, May 24, 2005, http://www.rigzone.com/.

[31] Catherine Bolton, "Browne in Talks as Field Is Reviewed," *Moscow Times,* May 30, 2005.

[32] Kazinform, "KazMunaiGas Considers the Possibility to Pipe Gas to China," Astana, February 5, 2005.

[33] David R. Sands, "U.S. Backs Kazakhstan-China Gas Pipeline," *Washington Times*, March 16, 2005.

[34] U.S. Environmental Protection Agency, *Reducing Methane Emissions from Coal Mines in China: The Potential for Coalbed Methane Development (Public Review Draft),* EPA 430-R-96-005, June 1996, p. 2-1. The EPA's central interest rested in how to reduce methane emissions from Chinese coal mines.

[35] Liu Honglin, Liu Hongjian, and Wang Hongyan, "China Has Good CBM Prospects but Few Commercial Projects," *Oil and Gas Journal* 102, no. 46 (December 13, 2004): 44.

[36] The state-owned China United Coalbed Methane Corporation was established in 1996. Several joint ventures have been set up with foreign companies, including Texaco China BV and Greka Energy Corporation, and a goal of 70 bcm of methane has been set for 2020, and five times that amount by 2020. See EIA, *An Energy Overview of the People's Republic of China*, p. 11.

[37] Yuri F. Makogon, Taras Y. Makogon, Stephen A. Holditch, "Russian Field Illustrates Gas-Hydrate Production," *Oil and Gas Journal* 103, no. 6 (February 7, 2005): 45.

[38] "Legislature Passes Renewable Energy Law," *China View,* February 28, 2005, http://www.chinaview.cn. "News Brief/China," in Platts *Oilgram News* 83, no.41 (March 2, 2005): 6, quoted the Xinhua News Agency in stating that renewable energy consumption accounted for 3 percent in 2004.

[39] "Legislature Passes Renewable Energy Bill," *China View,* March 1, 2005, http:// www.Chinaview.cn.

[40] Ibid., March 1, 2005.

[41] "News Brief/China."

[42] "Building Energy Conservation on Agenda," *China Economic Net,* February 23, 2005, http://www.ce.cn.

[43] Timothy Gardner, "Many Nations Have Great Potential for Renewable Energy," *The Progress Report*, www.progress.org/2005/energy46.htm.

[44] "China to Tap Hydropower Use," *China Economic Net,* May 24, 2005, http://en.ce.cn/national/government.

[45] U.S. Department of Energy, "An Energy Overview of the People's Republic of China," 2005.

[46] Fiona Harvey, "High Oil Price Drives Users to Other Fuels," *Financial Times,* March 16, 2005.

[47] China enjoys the unwelcome reputation of securing access to the latest technology by any means necessary.

[48] States News Service, "Beijing Cleans Air for 2008 Olympics with US Help," April 15, 2005.

CHAPTER EIGHT

A QUESTION AND AN ANSWER

SUSTAINABILITY IS THE QUESTION

The key word today is "sustainability." Is oil demand growth in China—and India—sustainable?[1] Is oil production and export growth in Russia sustainable? Will the oil-exporting countries of the Persian Gulf and OPEC in general be willing and able to successfully develop and sustain producing capacity in excess of world oil demand?

Beyond China's current and future needs for oil and natural gas imports, China today has a voracious appetite for steel, copper, and cement, among other commodities. In sum, China imports so that it can export. Economic expansion in China essentially reflects, and will continue to reflect, its ability to export goods to world markets and the continued growth of the world demand for their exports. Today it is textiles, tomorrow it could be automobiles, as China puts its natural advantage—low-cost labor and an artificially undervalued local currency—to work.

There are and will continue to be objections to how China uses these advantages. Any slowdown in Chinese exports, for whatever the reason, likely would be reflected in reduced commodity imports, quite possibly including oil. Countries that have thrived on selling to China would not be insulated from that slowdown. If oil requirements declined, would the decreasing oil demand be sufficient to depress prices? Would OPEC respond by reducing production and exports? Too many variables are at play to permit any acceptable assessment. This prospect adds uncertainty to the world oil market and, in turn, supports continued price volatility.

A somewhat unusual press conference was held at CIA headquarters in January 2005. The press conference was called to bring attention to a newly released unclassified report entitled *Mapping the Global Future*, prepared by the National Intelligence Council.[2]

Holding a press conference at the CIA is a somewhat unusual event, but perhaps no more unusual than certain of the findings contained in the report.

- The world of 2020 is likely to be one in which Asia is the main engine of the global economy, where China and India are major powers.
- The likely emergence of China and India as new major global players, similar to the advent of a powerful United States in the early twentieth century, will transform the geopolitical landscape, with impacts potentially as dramatic as those in the previous two centuries.
- By 2020, China's GDP—the total value of goods and services—will be greater than that of any Western country except the United States. India's GDP will have overtaken, or will be close to overtaking, European economies.
- Led by China and India, Asia looks set to displace Western countries as the focus of international economic dynamism—provided Asia's rapid economic growth continues.

Although the report is confident of China's continued strength on the world stage, Asia's rapid economic growth is not necessarily a given. This uncertainty adds uncertainty to the world oil market and in turn supports continued price volatility.

For example, the following problems confronting China today will not be easily resolved:

- Growing dependence on imported oil and, soon after, on imported natural gas;
- Increasing water shortage;
- Mass migration from rural to urban areas;
- An aging population that raises the question whether China will grow old before it grows rich;
- A rising possibility of agricultural shortages, leading to dependence on grain imports.

China consumes 10 times more water per capita than developed economies. It is accepted that two-thirds of the cities in China are short of water, that 90 percent of the rivers are polluted, and that 20 percent of available water supplies are lost through leakage.

Labor migration is not unusual—it is a natural occurrence. It is the scale of labor migration in China that is unequaled, perhaps the largest mass migration in history. In recent years, Chinese cities have absorbed at least 114 million rural workers, and they are expected to see an influx of another 250 million to 300 million in the next few decades.[3]

The rationale for the shift from the land to the city came out of the 10th National People's Congress and the 10th Chinese People's Political Consultative Conference. "Let the farmers move to cities to develop labor-intensive industries," and "to help farmers get rich, it is necessary to reduce the number of farmers by shifting surplus rural labor in a big way."[4]

The problem is this: a large rural population versus limited farmland. Taking farmers off the land will allow large-scale intensive land operations that may reduce the cost of agricultural products. For some regions, moving farmers to the city becomes the only way for farmers to increase income and for officials to point to higher industrial performance. But again, how can the government acceptably address the issues of pensions, public health, and social security costs?

China faces a period of rapid aging that will outpace the aging of most of the world's population. Between 2010 and 2040, the proportion of people aged 65 and older will increase from 7 percent to 25 percent.[5] This dramatic demographic shift gives rise to a number of questions, including how to provide and sustain a sufficient retirement income and a minimum level of health care for the elderly, who will number more than 332 million in 2050.

How does China compare with the United States, for example? According to the U.S. Census Bureau, by 2025 less than 20 percent of the U.S. population will be younger than 15, just slightly less than today.[6] The number of those 65 and older, however, will expand only from 12.4 percent of the population to 18.2 percent.

Not only must China confront all the problems associated with aging—aging in China has been described as faster than any other country in history—but its population growth will steadily decline as well, then turn negative.[7] As a result, China's population in 2050 will have fallen below the 2025 level, according to the United Nations' latest *World Population Prospects.*[8]

This demographic shift, which began in 2004, can be traced to 20 years of family planning—the "one-child" admonition. As a result, the number of people entering the labor force is going to decline for the next 15 years.[9]

China's population decline in turn will allow the population of India to overtake that of China before 2020, according to the UN report. As a result, India will come under greater pressure during the coming years to line up energy supplies in amounts sufficient to support the needs of 1.4 billion people in 2025, and about 1.6 billion in 2050.

Lastly, for a country with 1.3 billion people, the question of self-sufficiency in grains is of the utmost importance in that food consumption is mainly concentrated on grain products. Although national advisers have concluded that current grain reserves are sufficient to cover the immediate needs, the future in their judgment does not look bright.[10] The advisers further noted that it is difficult to improve grain output with more and more land being taken out of production as industrialization and urbanization take their toll. At the same time, a rising population and an improvement in living standards, leading to consumption growth, further support the observation that the future food supply situation cannot be regarded with much optimism.

Beyond these clearly definable issues, another constraint lurks in the background, a constraint now clearly recognized as a threat to China—HIV/AIDS.[11] Formidable challenges lie ahead if China is to successfully combat and contain its spread.

IT'S NOT JUST CHINA

Of particular importance for OPEC member-countries, will the economies of China and India continue to expand at current rates in the coming years or will they stabilize at lower levels? Will expansion of non-OPEC oil supply limit OPEC exports or, conversely, will growth in non-OPEC supply slow, then possibly plateau, thus leaving a larger market share to OPEC?

Will OPEC be able to expand its spare producing capacity in the coming years both to meet growing demand and to offset any surprises? Or, will spare capacity availability lag demand expansion, thus supporting continued tight market conditions?

Spare capacity contains the risk of translating into lower prices, if use of that capacity was not controlled in some fashion. In other words, unused spare capacity is a frozen asset, but putting that capacity to work, in the absence of offsetting demand, to realize a return on investment or in response to budgetary requirements, likely would lead to reduced prices.

Asia has been a driver of world oil demand during the past 10 years, if not longer, and accounted for about 40 percent of global oil consumption of some 82 million b/d in 2004. Lacking its own resources, Asia has had to turn to imports to cover the incremental growth in oil demand.[12] Japan is by far the largest Asian importer of oil, followed by South Korea, China, and India. Their aggregative oil imports in November 2004 averaged almost 11.8 million b/d, approximating oil imports by the United States. The Middle East has been and continues to be by far the dominant supplier to these countries.

China, in its search for access to oil outside the country, appears determined not to repeat the approach taken by Japan in its own search for foreign oil supplies. The key difference is very much apparent. China's efforts are more directed toward acquiring what has already been developed, whereas Japan, through the Japan National Oil Company (JNOC), endeavored to explore, develop, and export from Japanese-owned and operated fields abroad. The goal set for such fields was to provide some 30 percent of the country's crude oil imports.[13] But JNOC, some 35 years after its establishment, is being dismantled, having failed to attain any acceptable degree of success.

The IEA early on recognized not only that China is the newest major player in the global energy system but that other players will have to make room for it. The IEA summed up China's actions in this way:

> Aware of its growing dependency on imported energy, China seeks a more prominent position in the existing global system of energy production and trade. Where it can, China seeks to open new connections in global markets. Increasingly, external energy policies are entwined in foreign economic and security policies in general.[14]

China, because of its voracious appetite for oil, has become part of the "new game" now defining the world oil industry. Prices have risen to a higher level and, barring any unforeseen circumstances, are judged to hold there for some time. It is not just China that requires increasing volumes of imported oil and natural gas to satisfy demand—rather, it is the Asia-Pacific region and developing countries in general. Moreover, the expanding role of NOCs of importer nations further complicates oil supply and demand issues. These companies, by their very nature, can offer terms far more acceptable to supplier nations than can IOCs. In these instances energy requirements are often difficult to distinguish from political goals.

EXPLORING THE FUTURE

Someone once observed that time is the greatest innovator. Perhaps time will give the world an acceptable substitute for oil. Will it be the hydrogen fuel cell, regarded today as the most likely replacement? Again, only time will tell.

Shell International is noted for its innovative use of scenario planning as a tool for helping managers plan for the future or rather for different possible futures. In its *Energy Needs, Choices and Possibilities*, Shell foresaw China's potential future development in possibly the following way:

> By 2025 China, with huge and growing vehicle use, faces an unacceptable dependence on oil imports. Unease about the sustainability of regional gas resources and fears about the reliability of external gas, suppliers push towards the use of indigenous coal. But this is becoming logistically and environmentally problematic. Land scarcity limits biofuel opportunities. . . . China is able to make use of (technological advances) to extract methane and hydrogen directly from its coal resources, allowing it to move energy by pipeline rather than thousands of trains. Once fuel cells take off in OECD countries, China starts to develop a transport and power system around cost effective fuel cells fuelled by indigenous methane and hydrogen.[15]

Will China respond to concerns about rising import reliance and the loss of energy independence in the same way the United States has—that is, showing much concern but taking little action? Could access to Western technology provide China by the year 2025 the means to move to a methane- or hydrogen-based economy? Might China's economic growth slow down—consciously or otherwise—in the coming years, to the extent that the fears of today and tomorrow are never realized?

Only time will tell.

Notes

[1] China sold in excess of 5 million motor vehicles in 2004 and is expected to sell more than 6 million in 2005, which goal, if realized, would support at least absolute increments in oil imports during the year. See "China to Produce 6 Million Cars in 2005," *China View,* February 5, 2005, http://www.chinaview.cn.

[2] National Intelligence Council, *Mapping the Global Future: A Report of the National Intelligence Council's 2020 Project,* December 2004. This is the third unclassified report released by the council in recent years that takes a long-term view of the future.

[3] Minxin Pei, writing in a *Foreign Policy* Special Report, "China Rising," January/February 2005. See "Dangerous Denials," p. 56.

[4] "China Should Prevent Latin American Pitfall When Shifting Rural Labor Force," *China Economic Net,* March 15, 2005.

[5] Robert Stowe England, *Aging China: The Demographic Challenge to China's Economic Prospects* (Westport, Conn.: Praeger/CSIS, February 2005). The author concludes that demographics will probably affect China's economic development and reform efforts—and its role on the global state—more than any other factor.

[6] Jonathan Weisman, "A Glimpse of Older America," *Washington Post,* May 22, 2005, p.1. The larger share (25.7 percent) of U.S. population in 2005 will be those in the 25 to 44 years age bracket, with that share down a bit from the current 28.2 percent.

[7] "Fewer Japanese," *Financial Times,* February 23, 2005.

[8] "India's Population to Outstrip China by 2030," *Financial Times,* February 24, 2005. China is not alone in facing a population decline. Japanese experts are predicting that the first decline in the country's population will occur in 2006. See "Fewer Japanese," *Financial Times,* February 23, 2005.

[9] Jim Yardley and David Barbhoza, "Help Wanted: China Finds Itself with a Labor Shortage," *New York Times,* April 3, 2005, p.4.

[10] "Grain Self-Sufficiency Still Key for Nation," *China Economic Net,* March 8, 2005, http://www.ce.cn.

[11] Bates Gill, J. Stephen Morrison, and Drew Thompson, *Defusing China's Time Bomb,* Report of the CSIS HIV/AIDs Delegation to China, April 13–18, 2004 (Washington, D.C.: CSIS, June 2004).

[12] Shiva Lingam, Vandana Hari, Kate Dourian, Alex Lawler, "Producers Court Asian Buyers in New Delhi," Platts *Oilgram News* 83, no. 4 (January 6, 2005): 1. During the first week in January 2005, India hosted a roundtable between oil exporters (Saudi Arabia, Iran, Kuwait, United Arab Emirates, Oman, and Qatar) and major Asian oil importing countries (China, Japan, Korea, and Malaysia), making a pitch for long-term crude oil supply contracts and mutual investments in upstream by consumers and downstream by producers to enhance oil market stability and security. (See "India Calls for Developing Asian Oil Market," *China Economic Net*, January 6, 2005, which cited Xinhuanet as its source.)

[13] FACTS Inc., "Japan's Energy Policy and Energy Security," January 2005.

[14] IEA, *China's Worldwide Quest for Energy Security.*

[15] Shell International Limited, Global Business Environment, *Energy Needs, Choices and Possibilities: Scenarios to 2050* (London: Shell International Limited, 2001), pp. 50–52.

INDEX

Page numbers followed by the letter t *refer to tables.*

ABOUT THE AUTHOR

Robert E. Ebel is chairman of the CSIS Energy Program, where he provides analysis on world oil and energy issues, with particular emphasis on the former Soviet Union and the Persian Gulf. He is also codirector of the Caspian Sea Oil Study Group and the Oil Markets Study Group. In addition, he was project director for a series of nuclear-related reports, including the Global Nuclear Materials Management Project (published January 2000), and for the three-volume CSIS report, *The Geopolitics of Energy into the 21st Century,* cochaired by Senator Sam Nunn and Dr. James Schlesinger (November 2000).

Mr. Ebel served with the CIA for 11 years and spent more than 7 years with the Office of Oil and Gas in the Department of the Interior. He also served for some 14 years as vice president, international affairs, at ENSERCH Corporation, advising the corporation and its subsidiaries on international issues relevant to day-to-day operations. He has traveled widely in the former Soviet Union, in 1960 served as a member of the first U.S. oil delegation to visit that country, and in 1970 was in the first group of Americans to inspect the new oil fields of Western Siberia. In 1997, Mr. Ebel led an International Energy Agency team examining the oil and gas sector of Turkmenistan and Uzbekistan. In 2002, he participated in the Sudanese peace talks, held in Machakos, Kenya, and in 2002–2003, he worked with a group of former Iraqi oil officials, under the Department of State's Future of Iraq Project, to produce an assessment of the Iraqi oil sector.

Mr. Ebel is a past chairman of the Washington Export Council and a past member of the board of American Near East Refugee Aid. His books include *The Petroleum Industry of the Soviet Union* (1961),

Communist Trade in Oil and Gas (1970), *Energy Choices in Russia* (1994), and *Energy Choices in the Near Abroad* (1997); he was coeditor, with Rajan Menon, of *Energy and Conflict in Central Asia and the Caucasus* (2000) and editor of *Caspian Oil Windfalls* (2003). Mr. Ebel is a frequent commentator on national and international radio and television, and his views on energy issues appear regularly in newspapers here and abroad. He holds an M.A. in international relations from the Maxwell School at Syracuse University and a B.S. in petroleum geology from Texas Tech. In 2002, he received the Department of State's Distinguished Public Service Award.